New Energy from Old Buildings

The Preservation Press
National Trust for Historic Preservation
1785 Massachusetts Avenue, N.W.
Washington, D.C. 20036

The National Trust for Historic Preservation is the only private, nonprofit organization chartered by Congress to encourage public participation in the preservation of sites, buildings and objects significant in American history and culture. Support is provided by membership dues, endowment funds, contributions and grants from federal agencies, including the U.S. Department of the Interior, under provisions of the National Historic Preservation Act of 1966.

This book was supported in part by a grant from the U.S. Department of Energy, Office of Conservation and Solar Energy.

The opinions expressed here are not necessarily those of the National Trust for Historic Preservation.

Library of Congress Cataloging in Publication Data

New energy from old buildings.

Bibliography: p. 193-96.
Includes index.
1. Architecture and energy conservation—United States. 2. Architecture—United States—Conservation and restoration. I. Maddex, Diane. II. National Trust for Historic Preservation in the United States.
NA2542.3.N49 720'.28'8 81-8516
ISBN 0-89133-095-X AACR2

Edited by Diane Maddex, The Preservation Press
Designed by Watermark Design, Alexandria, Va.
Composed in Times Roman and ITC Italia by Harlowe Typography, Washington, D.C.
Printed by Collins Lithographing and Printing Company, Baltimore

Cover photo © Philip Trager 1977. Columbia Congregational Church (1832), Columbia, Conn.

New Energy from Old Buildings

National Trust
for Historic Preservation

Energy—or the lack of it—has shaped the nation's buildings from time immemorial. From the solar-oriented puebloes of native Americans to the half-buried sodbuster homes of the Midwest, from New England's saltboxes to Charleston's breezy piazzas, much of America's architectural evolution documents a struggle to defeat the less pleasant aspects of climate and environment *without* energy as an ally. . . . But with the onset of the energy crisis, designers have gradually become more aware of their forebears' struggles, and their solutions.

Kevin W. Green
Research and Design, Summer 1980

The solar-oriented pueblos of Acoma, N.M., date from 1200 and may be the oldest continuously occupied settlement in the United States. (Photo: Joshua Freiwald)

This 19th-century sod dugout in Norton County, Kans., was half-buried for protection from extremes of heat and cold. (Photo: Kansas State Historical Society)

New England's saltboxes such as the Jethro Coffin House (c. 1686), Nantucket, Mass., provided protection from the wind. (Photo: Cortlandt Hubbard, Historic American Buildings Survey)

Shaded deep porches and piazzas as used on the Charles Edmonston House (1828), Charleston, S.C., were designed to capture offshore breezes. (Photo: Louis I. Schwartz, Historic American Buildings Survey)

Contents

3. Alternative Energy Sources for Old Buildings

4. Afterword

Acknowledgments

Most of the papers in this book were presented originally at the symposium "Preservation: Reusing America's Energy," held in May 1980 under the cosponsorship of the National Trust for Historic Preservation and the Smithsonian Institution Resident Associate Program. The symposium was held during the 1980 National Historic Preservation Week, which also featured the energy conservation theme.

The following papers were not presented during the symposium and are included here with the permission of their authors and sponsors, which the National Trust for Historic Preservation gratefully acknowledges:

"Energy Conservation: Preservation's Windfall," by Neal R. Peirce, presented at the 1980 annual conference of the Maryland Historical Trust.

"An Old-House Conservation Strategy," by Nathaniel Palmer Neblett, AIA, adapted from the publication *Energy Conservation in Historic Homes,* © 1980 Historic House Association of America.

"How to Save Energy in an Old House: A Chart," by Douglas C. Peterson, published originally in *Historic Preservation,* March-April 1979, the magazine of the National Trust for Historic Preservation.

"Energy Guidelines for an Inner-City Neighborhood," by Travis L. Price III, contributed by the author.

"Economic Considerations of Solar Systems," by James S. Moore, Jr., P.E., adapted from a presentation at the 1980 annual conference of the Maryland Historical Trust.

Grateful acknowledgment is made for permission to reprint the following copyrighted material:

From "Energy Conscious Redesign" by Kevin W. Green, *Research and Design,* Summer 1980. Reprinted by permission of the author and the AIA Research Corporation.

From *The Social Life of Small Urban Spaces* by William H. Whyte. Copyright © 1980 by William H. Whyte. Reprinted by permission of the author and the Conservation Foundation.

From the *Washington Post,* May 6, 1980, by Dick Dabney. Reprinted by permission.

From *An American Architecture* by Frank Lloyd Wright. Copyright © 1955. Reprinted by permission of the publisher, Horizon Press, New York.

The National Trust for Historic Preservation also wishes to thank the following persons:

For assistance with Preservation Week 1980 activities, administered by the National Trust Office of Public Affairs, Lyn E. Snoddon, vice president: Tom Tatum and Rhett Turnipseed of the U.S. Department of Energy.

For the symposium "Preservation: Reusing America's Energy": the symposium organizing committee, directed by Janet Solinger of the Smithsonian Institution, including members Michael Alin and Ed Gallagher of the Smithsonian Institution; William I. Whiddon and Phillip Johnson of Booz, Allen and Hamilton; and Lyn E. Snoddon and Tom Donia of the National Trust.

For publication of this book: Terry B. Morton, vice president and publisher, the Preservation Press; Lee Ann Kinzer and Melanie Dzwonchyk, for editorial services; Sarah C. Gleason, for photo research; Baird M. Smith, AIA, and Jean Travers of the U.S. Department of the Interior, for editorial consultation; and Ellen Coxe of the Maryland Historical Trust and James C. Massey of the Historic House Association of America, for permission to publish their conference papers and publication respectively.

Diane Maddex
Senior Editor, Preservation Books
The Preservation Press

Foreword

Silver Dollar Saloon (1893), Nashville, Tenn., headquarters of Historic Nashville. The banner—proclaiming that 42,200 gallons of gasoline would be required to replace the building—was part of a National Trust 1980 Preservation Week program to promote energy conservation through reuse of historic buildings. (Photo: National Trust)

The idea of preserving old buildings is hardly new. In the United States preservation as a formal endeavor dates to the mid-19th century; it is, of course, much older in other parts of the world. But it was not until 1949, when Congress chartered the National Trust for Historic Preservation, that there was a national organization devoted to that single concept.

Since that time, the idea of preservation has spread far and reached a great many people. The technology of preservation has improved. The reasons for preservation have increased. When the National Trust began, preservationists were a relatively small group concerned mainly with the oldest, most historic sites that were threatened by neglect or demolition. By the 1960s preservationists were fighting the large-scale destruction of historic buildings carried out in the name of urban renewal.

But it was not until the 1970s that preservation began to be associated with other social and economic goals, including the revitalization of urban areas, the protection of valuable farmland, increased employment opportunities, safe and adequate housing and the reduction of wasteful consumption in all segments of society.

To address such problems through preservation, it became necessary to form new alliances, build new coalitions and seek new allies. Architects and planners began to show a greater interest in preservation; business owners and foundation heads started to support established preservation programs and initiated their own. City administrators and officials took note; politicians, bankers, farmers, educators and a cross section of the nation became preservationists. By 1980 there was a formal alliance between preservation and natural area conservation, with organizations in those two fields cooperating on legislation, federal funding and other issues that affected both.

Despite this history of coalition building, until recently only a few dreamers would have predicted that a relationship would develop between preservation and energy conservation. Energy questions surfaced with urgency following the Arab oil embargo and subsequent rapidly rising energy costs in the 1970s; preservation, it seemed, would have to take a back seat to other more pressing national concerns. Some architects knew empirically that older buildings were more energy efficient than the post-World War II generation of structures, but they did not have broad statistical proof.

By the beginning of the 1980s there was such proof. A 1977 study of office buildings in New York City [see page 63] found that buildings erected after 1940 consumed more energy per square foot than those built in the previous four decades and before—an average of 25 percent more, and greater than twice as much in individual cases. The study showed that the oldest buildings generally used the least energy. This confirmed the view of many that America had been constructing energy dinosaurs during the postwar era of cheap and abundant fuel.

Following on the heels of this study was another one that, for the first time, quantified the amount of energy required to demolish an old building and replace it with a new one of similar size and function. The startling conclusion was that every eight bricks in place held embodied in them the Btu equivalent of a gallon of gasoline. In virtually every specific case to which these newly devised formulas were applied, the argument could be made and proved with numbers that a replacement program would be energy wasteful.

This book, which is based on a national symposium on energy conservation and its relationship to preservation, summarizes and explains aspects of those studies. It suggests new directions that the nation can take both to cut energy consumption and retain tangible elements of its rich heritage.

Most important, it reaffirms that preservation is not just a mechanical or legal or economic process. It is an ethic. The simple philosophy of reusing what is best from the past and rejecting the throwaway mentality is the same ethic behind protecting wildlife, guarding the beauty of fragile natural areas and saving gasoline, fuel oil and electricity.

The fact that preservation conserves energy must now be taken to our legislators, our corporate leaders and our opinion molders. It must become the foundation for national policy on the built environment. We must find, highlight and change the laws, practices and misconceptions that have led us as a nation to treat buildings as simply more disposable items, rather than the capital assets that they are.

Neither preservation nor conservation, nor energy issues, nor the future of our cities and towns and rural areas is a simple matter, unrelated to the others. Together they all determine the quality of life in America today, and the plans and policies now being set in these interdependent areas will determine the quality of life in the future.

Michael L. Ainslie, *President*
National Trust for Historic Preservation

Introduction

Preserving History and Saving Energy: Two Sides of the Same Coin

John C. Sawhill

The energy problem, perhaps more than any other influence, has taught us the limitations of our resources and the urgent need to conserve, protect and defend our national patrimony. The values that historic preservation has long represented are gradually becoming the norm. Although change is still a hallmark of our fast-paced society, civic progress is no longer so intimately identified with the wrecker's ball. And it is fair to say that communities that have paid heed to preservation are more interesting and pleasant places in which to live than are those that have permitted, or even abetted, the destruction of distinctive old buildings and neighborhoods.

But the case for preservation no longer rests solely on aesthetics; the preservation movement embraces more than a handful of antiquarians, "history buffs" and public-spirited citizens. Business people and homeowners alike are coming to recognize that rehabilitating old buildings for new uses is often less costly than new construction, and retailers know that a historically or architecturally interesting environment is an irresistible attraction to tourists and shoppers.

Recently, I visited the site of the 1982 World's Fair in Knoxville, Tenn. The theme of that fair will be energy. Situated among the ultramodern exposition halls being erected are other structures being preserved because of their architectural and historical significance. The U.S. Department of Energy is supporting this worthy effort. A number of residences, a railroad station and a candy factory will be restored and will demonstrate that energy efficiency and preservation go hand in hand.

Opposite: Monadnock Block (1891, Burnham and Root; 1893, Holabird and Roche), Chicago, c. 1900, to be restored with energy-saving measures. (Photo: Commission on Chicago Historical and Architectural Landmarks)

End of the Cheap-Energy Era

In the days of cheap energy, it might have made a kind of short-sighted economic sense to destroy old structures, but that era ended forever with the oil embargo of 1973. In that year, the United States imported one-third of its petroleum. Today, the figure is almost 50 percent. We are spending more than $10 million an hour on imported oil—$90 billion a year. If such expenditures continue at the present rate, in less than 12 years we will have exported cash equal to the trading value of all the stocks listed on the New York Stock Exchange. The headlines remind us daily that we are highly vulnerable to disruptions in supply occasioned by acts of war, boycott, terrorism or revolution.

The Carter administration sought from its beginning to diminish the nation's reliance on imported oil. A multifaceted program to expand production of energy of all kinds has been structured, and we are working closely with our friends and allies abroad to stabilize the world petroleum market. But conservation must be the keystone in this structure. It is far cheaper to conserve a kilowatt of electricity or a gallon of gasoline than it is to produce one.

Studies by the National Academy of Sciences and others show that at least 25 percent of the energy that normally would be used in buildings over the next 10 years can be saved. Residential and commercial buildings consume approximately 38 percent of the energy expended in this country, and an estimated 41 percent of this energy—or the equivalent of 5.7 million barrels of oil per day—is wasted. It has been estimated that widespread retrofitting to make buildings more efficient could save 8 quadrillion Btu's of energy annually by 1990—the equivalent of discovering two new oil fields the size of Alaska's North Slope.

It is true that if an old building is replaced with a modern, tightly designed, energy-efficient structure, there will be an immediate net savings in operational costs. These savings would seem to argue against preservation in strict dollars-and-cents terms. But what seems obvious at first glance is not necessarily so. The fact is that an existing building represents a certain repository of value. It took energy, materials and human labor to put it up, and it would take energy, machines and labor to tear it down.

Energy Investment of Buildings

The value of materials in place always should be considered before demolition proceeds. Take the example of a five-ton steel girder delivered to a construction site. The energy invested in processing and fabricating the girder is 257 million Btu's; transporting it to the construction site and installing it might require 13 million Btu's. By leaving it in place, the contractor would eliminate the consumption of a total of 270 million Btu's—the amount of energy in 2,000 gallons of gasoline. And the price of the girder would be saved.

America's building stock is one of its most significant energy investments. Replacing all the existing buildings in the United States would require the world's entire energy output for one year—approximately 200 quadrillion Btu's. Preserving structures and using them as long as possible thus is clearly justified. In fact, the Advisory Council on Historic Preservation studied three major preservation projects and found that rehabilitating existing structures could save 50 percent or more of the energy needed for new construction [see discussion, pages 103-19].

The Department of Energy is trying to extend the pioneering work of the council. In cooperation with the U.S. Department of Housing and Urban Development, it has funded a study of the Manchester Neighborhood in Pittsburgh to determine the amount of energy that can be saved by revitalizing an entire city block. The project sponsors believe that a neighborhood such as this one can be renovated and operated for 40 percent less energy and water than could a suburban subdivision with an equal number of housing units. Encouraging people to live in the city also encourages more intensive use of mass transit. The study results will be disseminated, and workbooks and materials will be prepared for distribution to

Top: Three of seven houses bordering the site of the 1982 World's Fair (International Energy Exposition) in Knoxville, Tenn., that were saved for use during fair activities. Forming the boundary of the Fort Saunders Historic District, the houses are an important link between the neighborhood and the new development.

Bottom: Houses following rehabilitation. Through a cooperative local effort and with financing from the National Trust, the houses will be resold after the fair by Knoxville Heritage. (Photos: Knoxville News Sentinel*)*

Louisville and Nashville Passenger Station (1904), Knoxville, Tenn., which will be used as offices and a restaurant during the 1982 World's Fair and after the fair as an exhibition center. (Photo: Ron Childress)

urban planners, architects, community leaders, lending institutions and homeowners to help them choose the best ways to rehabilitate old housing [see pages 93-99].

In addition, the Department of Energy has been working with 60 cities throughout the country to establish community-wide energy-planning processes. The department hopes to convince Congress to provide funding directly to local governments to further encourage innovative conservation planning at the community level. I hope that some of these planning and demonstration grants can focus on preservation and its relationship to energy conservation.

The link between the preservation and restoration of buildings and energy conservation is strong. The era of cheap energy led to urban sprawl and a lifestyle dependent on the passenger car. The number of cars increased 178 percent between 1950 and 1977, while mass transit systems declined in performance and use. The preservation and restoration of our cities and total community planning can change that and benefit all of us, both by increasing the efficiency of energy used in buildings and making mass transportation more accessible and appealing.

Many communities today appreciate the relationship between preservation and conservation. Yet it must be recognized that some old buildings may never be quite as energy efficient as newer structures. Even so, the lower capital cost they represent will usually offset for many years any operating energy advantage that tearing them down and replacing them might afford. On the other hand, many old buildings are inherently more efficient than any newer designs that would be economical to build. The massive walls of many old buildings, for example, provide good insulation and high thermal inertia; they absorb heat and release it slowly to moderate any day-night temperature differences.

In general, the least energy-efficient structures are those built between 1941

Littlefield and Steer Candy Company (1914), Knoxville, Tenn., used most recently as a department store warehouse. Located on the 1982 World's Fair site, the building will house public and employee restaurants that will become permanent facilities after the fair. (Photo: Ron Childress)

and 1970. Newer buildings often are more efficient because high energy costs have encouraged better insulation. Pre-1941 structures typically use less energy because they maximized natural sources of heating, lighting and ventilation.

One caveat should be mentioned here: Owners of historic buildings should limit retrofitting measures to those that achieve reasonable energy savings at reasonable costs with the least impact on the character of the building. Overzealous retrofitting introduces the risk of damage to distinctive features and can impair aesthetic values.

National Change of Attitude

It is highly gratifying that the necessity of energy conservation in the home is finally sinking in. The Opinion Research Corporation recently conducted a nationwide survey concluding that homeowners are deeply concerned about rising fuel prices; nearly half of those questioned said they had been adversely affected financially. Eighty-three percent said that they believed the energy efficiency of a house would affect its resale value, and 95 percent regarded energy conservation as an important consideration in purchasing a new home.

When questioned about how much they were willing to spend for a comprehensive insulation package that could save 25 percent of annual heating and cooling costs, responses varied from a low average of $1,200 in the South to a high average of more than $1,500 in the North-Central states. When asked whether energy-efficient residences should qualify for lower mortgage rates, 63 percent agreed that they should.

Some consumers are following up their beliefs with action. Thirteen percent of the nation's homeowners already have taken advantage of recently enacted federal tax credits for conservation and renewable energy sources. Over the past

two years, homeowners have made average investments of more than $700, with a total value of more than $7 billion. Such investments are being made primarily by higher-income families, but the Department of Energy is working with Congress to broaden the appeal of energy efficiency in homes. Legislation to provide a solar and conservation bank that would help low and moderate-income families retrofit their houses was passed recently.

The department also is taking direct action through its weatherization program to insulate the homes of the poor, elderly and handicapped. To date, more than 490,000 houses have been weatherized. This program is moving forward at a rate of 28,000 additional residences a month.

New Energy Conservation Directions

The Department of Energy has instituted a number of new programs to help homeowners conserve energy. It is sponsoring extensive research to develop better materials, methods and processes for retrofitting buildings, as well as for new construction. It is demonstrating an unusual heating and cooling system—the Annual Cycle Energy System (ACES)—capable of cutting residential fuel bills in half. In its first year of operation, an ACES model house saved 50 percent over the control house next door.

The ACES system features a swimming pool-sized tank of water in a basement of crawl space for use as an energy storage bin. The water in the tank stores summer heat, which is used to warm the house in the winter and to provide hot water. As heat is drawn from the water by a heat pump, it slowly turns to ice. In summer, chilled water from the melting ice is used for air conditioning.

Private industry, in partnership with the federal government, also is being encouraged to develop and commercialize new energy-saving technologies. The

Former Savannah Volunteer Guards (1893, William G. Preston), Savannah, Ga., converted for use as the Savannah College of Art and Design. The thick walls, tall windows and porches are energy-conserving features shared by many historic buildings. (Photo: Marcia Axtmann Smith)

Department of Energy has cooperated with a real estate developer near Los Angeles to demonstrate residences that attained fuel savings of 50 percent mainly through conservation techniques in comparison with the energy requirements of conventionally built houses. Another demonstration house, on Long Island, is expected to save 80 percent in energy use through a combination of conservation and passive solar features.

One of DOE's major new programs is the attempt actively to interest the electric utilities in residential energy conservation. The department is working with Congress to remove the barriers that prevent the utilities from playing their natural role as vendors of a broad range of conservation services. It also is working with utility executives and regulators to assure that investment opportunities in conservation will be given the same consideration as opportunities to increase generating capacity. The Department of Energy would be delighted to sponsor joint projects with utilities and the Advisory Council on Historic Preservation designed to achieve the dual objectives of conservation and preservation.

The Department of Energy is anxious to explore a variety of energy-conservation delivery systems. One might employ municipal agencies or nonprofit corporations to provide a range of technical and financial retrofitting services. Another envisions energy-service corporations that would take over the payment of a householder's utility bills and make their profit from savings produced by installation of conservation systems. Municipalities might provide tax benefits for such retrofitting or penalize the transfer of property that had not been rendered energy efficient.

The preservation and renovation of residential, commercial and industrial structures will itself be a major industry within a few years. The force behind this emerging industry is the philosophical transition away from a throwaway society based on an illusion that resources are unlimited—toward a new kind of civilization grounded in performance, balance and order. Through historic preservation's leadership, it will be made obvious to all that two vital social goals—energy conservation and historic preservation—are self-reinforcing. Each can assist the other in obvious and important ways. It behooves all of us, as preservationists and as energy consumers—to make this linkage better known. By working as allies, we will speed the day when our communities are once again healthy, efficient and attractive.

Energy Conservation: Preservation's Windfall

Neal R. Peirce

Throughout our history, Americans have been accustomed to an economy of bounty. We arrived on these shores to find a largely unsettled continent of immense challenge, a wilderness we thought demanded conquering, a broad frontier physically and in our minds and spirits. Through the years of westward expansion until the farthest West was reached, through the years of great industrial growth in the 19th and early 20th centuries in the North and Midwest, and through the phenomenal Sunbelt and suburban growth after World War II, we never sighted limits.

But now, chiefly because of energy, all that is changing. And it is not enough for local, state or national economic survival to make half-hearted reforms. Our national interest requires a total reorientation of our energy-consuming habits. We will have to proceed with a unified national effort in house and factory and office building weatherization, for more energy-efficient autos, radically reduced driving, industrial cogeneration of power, more coal, plus solar and other renewables; and a new conserving form of land usage.

The back-to-the-city movement, the entire urban rebirth we are witnessing today, has been influenced to substantial degree by these realities.

This is a particularly auspicious time for all those interested in city revitalization and preservation—a time, perhaps, of some new dangers for historic buildings and neighborhoods, but still, a time when the constellations are arranged in favor of preservation as rarely in the past.

Opposite: Glencoe Place row houses (1890s), Cincinnati, renovated by the Mount Auburn Good Housing Foundation to help retain neighborhood residents. (Photo: Carleton Knight III, National Trust)

And how things have changed!

A decade ago there was discouragement, near despair, about many of our center cities. Some city critics even took glee in this. A Columbia professor, Eugene Raskin, could then write in the *Reader's Digest* that cities "are physically obsolete, financially unworkable, crime-ridden, garbage-strewn, polluted, torn by racial conflicts, wallowing in welfare, unemployment, despair and corruption." He suggested that they were "unsalvageable" and "deserved extinction."

A Changed Time for the City

Today, even if cities continue to have problems of crime and joblessness and pollution—and they do—and those problems are often grim and intractable, the kind of picture Professor Raskin painted sounds simply absurd. Why? Let me mention just a few signs of the changed times.

We have witnessed a remarkable surge of downtown office building—up 50 percent over the course of the 1970s in cities of 150,000 people or more, according to the Urban Land Institute. The suburban office park setting for corporations has lost some of its allure; instead, the desire for eye-to-eye contact with peers, and for a central location, has caused corporations, law firms and financial institutions to seek out downtowns, underscoring a major, historic role of the city. Concurrently, business and tourism have spurred demand for centrally located facilities such as hotels and large meeting places, and thus helped to rekindle retailing in many center cities. Small-town retailing has been ravaged by the suburban malls, but larger cities are, on the whole, finding ways to adapt and survive.

Now, instead of downtown desolation, we see the 1980s as a decade in which the fight will be over who controls a suddenly

Harborplace (1980, Benjamin Thompson and Associates), Baltimore, developed by the Rouse Company as a key feature of the revitalization of the city's Inner Harbor. (Photo: The Rouse Company)

more valuable urban turf, with the actors ranging from opulent corporations to poor neighborhood residents.

Baltimore, for example, is in the midst of one of the world's most astounding urban revitalization projects. Charles Center and the Inner Harbor, with its many attractions capped off by Harborplace, are being looked to from coast to coast as examples of what long-range planning and vision can do for once seedy and neglected center cities.

James Rouse, developer of Harborplace, is full of stories about how financiers and retailers were simply incredulous when he tried to launch Faneuil Hall Marketplace in Boston and the Gallery at Market East in Philadelphia. But now, with those projects prospering and the almost unbelievable success of Harborplace, some critics are already picking at the concept. They are muttering that the concept of Rouse's so-called festival marketplaces will become routinized, that these will be the new Howard Johnsons of downtown U.S.A. Well, it is true that sterile thematic copy-catting is always a danger. Maybe there are limits to how many hanging ferns we can stand. But to have a ghost of succeeding, projects such as Rouse's must be carefully planned to fit into the fabric of the city. They cannot, like some great office or hotel megastructure, be plunked down anywhere on the urban landscape and be expected to succeed because of their moat-like security and total interior environments. They have to fit into the pedestrian flows of the city, they must represent contextual architecture of some excellence, they must show sensitivity to the surrounding cityscape in countless ways.

In his new book, *The Social Life of Small Urban Spaces* (Conservation Foundation, 1980), William H. Whyte catches this point quite elegantly:

> It is significant that cities doing best by their downtowns are the ones doing best at historic preservation

The rehabilitated Quincy Market (1826, Alexander Parris), Boston, transformed into Faneuil Hall Marketplace (1976, F.A. Stahl and Associates; Benjamin Thompson and Associates). (Photo: David Cubbage, the Rouse Company)

and reuse. Fine old buildings are worthwhile in their own right, but there is a greater benefit involved. They provide discipline. Architects and planners like a blank slate. They usually do their best work, however, when they don't have one! When they have to work with impossible lot lines and bits and pieces of space, beloved old eyesores, irrational street layouts, and other such constraints, they frequently produce the best of their new designs—and the most neighborly.

Richard Fleming, who left his job at the U.S. Department of Housing and Urban Development to become president of Downtown Denver, Inc., has asserted that "center city development interests are recognizing that 'successful downtown redevelopment' is not simply measured at the skyline but, equally important, at the ground level—in the streetscape, parks, plazas, storefronts, and facades that make up a downtown's public spaces."

Of course, that is not universally so. Massive architectural assaults are still being carried out on many of our center cities. I see them all the time during my travels. But note an interesting trend of the times: the general disappearance of a familiar phrase, "central business district." The insidiousness of that expression, suggesting that downtown exists for nothing except business was first brought home to me in an interview two years ago in Vienna with Victor Gruen, the great city planner. Gruen is no longer with us, and the world is a poorer place for it. But he would be pleased to hear, as I have on my travels, many people assaulting the CBD concept for the distortion of the true city it represents.

We are, I believe, witnessing a major, historic shift in architectural tastes. We were treated to several decades of the self-assured modern movement in architecture, which brought us the organic stone fortresses of Le Corbusier, the functional glass towers of Mies van der Rohe and the similar works of all their disciples. They seemed to equate social progress with technological progress and to reject history.

But as architecture critic Wolf Von

Ecker Drug Store (1885), Corning, N.Y., after it was restored as part of the town's Main Street revitalization effort. (Photo: Norman Mintz, Market Street Restoration Agency)

Ecker Drug Store during restoration, showing the removal of the siding that had covered the original facade. (Photo: Norman Mintz, Market Street Restoration Agency)

Eckardt noted in the *Washington Post*, "A destructive delusion is lifting from our cities. It is the delusion that we are living in a totally new age in which only the new is relevant." And "lifting the delusion," he wrote, is a "sudden historic awareness [that] has propelled the historic preservation movement to unprecedented importance." No longer, he suggested, is the chief fight over preserving certified landmark buildings. It has moved on to the perhaps grayer but equally important area of preserving the integrity of entire neighborhoods or city areas. It has infected architecture so that it is no longer considered bad taste for urban designers to relate to historic architectural styles. It is no longer said that the modern and the abstract are the only styles that can express the spirit of our time.

Shifting Demographics, New Lifestyles

There is another major factor that makes this such an auspicious moment for the forces of city revitalization and historic preservation. It is the shifting demographics and change in lifestyle preferences.

During the late 1970s we began to hear a lot about the back-to-the-city movement—the choice, especially among young professionals, of previously neglected city neighborhoods in which to live. Often, as we know, they restored old houses, an immense plus for the preservation movement. Many selected the inner city because it was closer to their places of work and because they had come to reject the uniformity and dullness of suburban settings.

Behind all this one can find the latter-day impact of the post-World War II baby boom. That huge population bump now is reaching its late 20s and early 30s, demanding an absoluetly unprecedented amount of new housing in this decade.

But this is no repetition of the marrying and child-begetting era after World War II. Today, people are spending more years living alone or with roommates or partners, delaying marriage, having fewer children, divorcing more, remarrying more slowly. Within this decade, according to a Harvard-MIT analysis, only 50 percent of American households will be headed by married couples, compared to 80 percent in 1950. By 1990, according to that study, there will be no typical type of household in this country: "The nuclear family consisting of mom, dad and the kids will no longer hold sway as it once did." Among other things, the proportion of women in the work force will continue to soar.

The impact of such shifts is already visible from the initial 1980 census returns. In city after city, while there often are population drops, sometimes rather sharp, the number of households is declining much less rapidly or in fact expanding. There are, in short, fewer people in more housing units. That may reflect more elderly people living alone or more fatherless families—a particularly severe problem in the minority community. But it also reflects houses subdivided for the convenience of young people who either prefer the city or find it more convenient in getting to work in those immense downtown office developments.

What this means for the 1980s, I believe,

is increasing demand for older city housing to be rehabilitated and recycled into more units, as well as to break the logjam holding back construction of new townhouses and apartments.

Saving Energy, Saving Buildings

Closely related to, and often behind, these developments representing a boon to city revitalization is the phenomenon of energy.

President Reagan has asserted that energy conservation is not as important an issue as we have come to think in the last several years, that all we need to do is to lift government regulations and drill or dig like hell for all the energy left beneath our soils. And then somehow we can return to the salad days of the 1950s and 1960s, to an era of a rapidly expanding national economy with energy no longer a real problem.

Many Americans wish fervently that this could be so. Maybe it can. But its likelihood of occurring is slim indeed. In 1973 this country was importing a very small portion of the oil it consumes and paying about $3 a barrel for it. Today, three times as much OPEC oil is imported, at prices ranging up to $35 a barrel. In 1971 the United States paid $4 billion for oil imports; in 1978, $40 billion; in 1979, $90 billion. Between now and 1985 we probably will pay $500 billion for foreign oil. That is more than half the total value of all stocks listed on the New York Stock Exchange—$900 billion. We will, in short, be sending up in smoke a huge portion of our accumulated wealth in a mere half decade. How we manage to do that and return to the easy energy times of yesteryear escapes me.

Concerns about energy prices, hopes of becoming less vulnerable to gasoline shortages and a genuine desire to assure that the country's valuable nonrenewable resources are recycled are at the root of the country's awakening interest in conserving old buildings along with other

Butler Square (1974, Miller, Hanson, Westerbeck, Bell), Minneapolis, located in the former Butler Brothers Warehouse (1906, Harry Wild Jones). In the building's adaptation to shops and offices, a skylight-topped atrium was created in part to open the interior to natural light. (Photo: Carleton Knight III, National Trust)

scarce commodities.

Existing buildings, as noted throughout this book, represent more than dollars and cents; they also are repositories of embodied energy. Many old buildings also are inherently more energy efficient than later designs, with their massive walls providing built-in insulation. My grandparents lived in such a house, and my travels have taken me to the arson-ravaged and abandoned areas of the South Bronx, where many similar sturdy old buildings with incredibly thick stone walls stand like ghostly sentinels. The thought constantly recurs: Why not, in an energy-scarce era, rehabilitate such buildings rather than letting them be levelled?

One of the most encouraging signs today in older neighborhoods such as Baltimore's is the vast number of traditional brick row houses that are being rehabilitated—and unquestionably causing great energy savings over demolition and then replacement elsewhere. What meets the eye may be just a part of the picture. Energy retrofitting of older small buildings can make them almost competitive with the best new construction, especially where one can easily seal around sills and windows and add energy insulation. But many smaller buildings have partial insulation that prevents complete weatherization.

The Office of Technology Assessment of Congress has a panel on energy for city buildings on which I have been participating. We have learned that large old buildings usually cannot be retrofitted to compete, in energy efficiency, with the best of new buildings. This is because truly energy-efficient large new structures almost always involve the clever use of building orientation to exploit daylight and the careful design of heating, cooling and ventilation systems to minimize utility costs. But assuming that older buildings are to remain, almost any of them can benefit from cost-effective retrofitting to provide 20 to 30 percent energy savings and occasionally up to 50 percent.

Federal Hill, Baltimore. The neighborhood, near the city's Inner Harbor, is being revived as part of the back-to-the-city revitalization movement. (Photo: Carleton Knight III, National Trust)

Vigorous outreach programs by utilities can induce homeowners to retrofit their houses—but often only if zero-interest weatherizing loans are offered. This is an experiment now being tried, with good results, by the Tennessee Valley Authority and by Oregon and California utilities. But it remains to be seen whether those programs can induce landlords to retrofit their buildings for energy savings, because landlords so easily pass utility costs on to their renters.

In city neighborhoods, energy conservation and retrofitting often are part of the house rehabilitation process. But conservation may compete for the homeowner's investment with better wiring and plumbing and facade improvements. For many neighborhoods, the OTA study has found that facade improvement receives the highest priority because it contributes most directly to upgrading the resale value of a property. That fact poses an interesting problem for preservationists if they wish to equate their cause with that of energy conservation.

In time, dozens if not hundreds of strategies, some quite complex, will have to be devised to achieve maximum energy savings through rehabilitation of old structures, in both cities and rural areas. But the practical problems should not be allowed to overwhelm the basic concept: that as a society learns to treasure its heritage and to recycle, adapt and reuse what was built before, then it will also, quite instinctively, make energy-saving choices. In my observation of the various themes that cities pick to anchor their revitalization strategies—from waterfronts to old markets, from community and fine arts to an industrial heritage—there seems to be a strong energy-saving component.

Building Coalitions for the 1980s

If energy concerns are to work for cities to the fullest possible degree, and to be harnessed among other things for the preservation cause, the 1980s will require the building of inventive and broad coalitions between preservationists and neighborhoods of all incomes and stripes, downtown businesses, universities and arts

Abandoned tenements in the South Bronx, N.Y., some of which are being rehabilitated through local "sweat equity" projects. (Photo: Mark Haven)

communities, industries, minorities, developers and all those concerned about the integrity of our rural communities and farmland preservation.

Land use is the issue that should draw together preservationists, environmentalists, minority communities and taxpayers concerned about ever-rising local levies.

A stark example comes from a study done five years ago by the Southeast Michigan Council of Governments to project the Detroit area's shape in the year 2000. It was estimated that a continuation of the recent pattern of rapid development into the suburban fringe would cause Detroit proper and its older, established suburbs to lose more than a third of their population by the end of the century. All growth would shift to the outlying suburbs. There would be an additional million cars on the region's roads, traveling an additional 40 million miles a day and wiping out every gallon of gas saved from more efficient autos. As much as 460 square miles of southeastern Michigan farmland would be lost to highways, subdivisions, shopping centers and industries by the year 2000. Detroit and its suburbs would abandon $2.4 billion worth of closer-in schools and have to replace them at even greater cost on the urban fringe.

This is a pattern of incredible waste—because all of the 700,000 new households anticipated in the Detroit area by the year 2000 could be accommodated within the network of already sewered city and suburban areas.

Today, a similar computer study might be somewhat less alarming; the worm has begun to turn toward somewhat less rapid outward expansion. But one thing remains clear: The cost of sprawl development, in dollars for new roads, sewers, schools and other public investments, in inner-city abandonment and neglect, in the destruction of once viable neighborhoods and in unnecessary energy use, are mind-boggling and obscene in a nation with the severe economic and energy supply problems that ours now faces.

Splurging Energy on Sprawl

Closely related is the issue of conserving agricultural land. At a breathtaking speed,

Countryside of Ashe County, N.C., part of the increasingly vulnerable rural landscape. (Photo: Michael Southern, North Carolina Division of Archives and History)

American farmlands have been sacrificed to subdivisions, shopping centers and factories. This lost land, situated often in fertile river valleys near our cities, is usually the country's most productive acreage. As land is devoured for development, more marginal, faraway western land must be substituted for agriculture. That land often requires irrigation, which requires energy to deliver. It also requires immense amounts of fertilizer, one of the most energy consumptive of all products.

Sprawl development is the flip side of the coin of urban decline. As one environmentalist recently put it, "For every block of abandoned houses in cities, there are probably three miles of choice farmlands out there in the countryside scraped and bulldozed away to build new homes. It takes away the farmlands, it wastes energy, and it creates a political force that sucks the cities dry."

The answer, of course, lies in much more dense development than has been built in the United States in many years. I believe that there is sufficient land, in most places, in these already sewered and serviced cities, suburbs and towns, places

that already have their schools, streets and fire services in place, to provide housing for the baby boom as it reaches the peak of its housing demand between now and the mid-1990s. The new housing must be in garden apartments, condominiums, townhouses and other forms of dense, transit-accessible housing. A lot of it may have to be built in empty inner-city and suburban lots, perhaps constructed directly over the factory and shopping center parking lots that consume so much of the space around us today.

A country in the fix we are in now should not try to resolve its housing needs by building more and more of the "American dream"—the free-standing house on its own lot in a low-density, totally auto-dependent suburb. Certainly the low-income residents of poor city neighborhoods cannot afford that kind of housing now. Indeed, a tinier and tinier percentage of Americans, even young professionals, find it affordable, given present and prospective construction costs.

Surburban housing typical of developments transforming rural areas. (Photo: From God's Own Junkyard, *1964, 1979, by Peter Blake. Courtesy Holt, Rinehart and Winston)*

We often assume that all housing developers are insensitive to these energy and space issues, anxious just to throw up their subdivisions on any cornfield outside of town and then let everybody else worry about the gasoline and public infrastructure costs.

My own thinking on the subject has been altered, and quite sharply, by a recent project entitled "Development Choices for the '80s." The research has been funded by HUD but carried out by the Urban Land Institute with a 50 percent participation by distinguished housing, real estate, banking and insurance executives from across the country. The message coming from that study is remarkable. It is that because of energy costs and housing inflation, much more dense and infill development, and much more mixed-use development of residential, office and commercial space, must come in this decade; that many developers want to move in this direction, and are doing so where they can. But one of the greatest obstacles is the mountain of local zoning laws that inhibit dense or mixed-use development.

Rethinking Old Patterns

Harold Jensen, the noted Chicago developer who is cochairman of the Development Choices Council, has some interesting observations on our country's prevalent single-use zoning pattern, with its tendency to segregate housing on one side of town and industry on the other. All this, he says, is really a throwback to the 1920s, when people were greatly concerned about having an offensive neighbor such as a rendering plant or a steel foundry. Now most such facilities have cleaned up their act or moved to the country. We have a body of environmental laws to protect us—laws that simply didn't exist a half century ago. Yet we are still stuck with the zoning laws, which badly need relaxing to achieve dense and mixed-use development— everything from mom-and-pop stores below residences to neighborhood pubs, from movie theaters in apartments to mixed shops, condominiums and recreation facilities, as well as housing that reduces home-to-work distances and thus energy use.

The related problem is the mindset of the public. Because housing has been about the only decent investment in recent years, people buy a home as much for investment as for shelter. The only secure investment, furthermore, has been thought to be the free-standing single-family house. This explains the view in many neighborhoods that townhouses or clustered housing or anything except single-family subdivisions is anathema, an attitude based on the theory that somehow they would reduce the value of one's own single-family home investment.

Think of the coalition opportunities all of this opens to preservationists and to those who want to do something about our country's energy problems. Think of some of the friends you can make among developers if you show a strong interest and join them before city councils in fighting to make pieces of already serviced city or suburban land available for housing of at least medium density. Or among minorities and the poor if you fight to make some of that housing available to them. Or how environmentalists and preservationists can cooperate on this whole set of issues.

The idea is not to choke off development and economic growth, but to encourage it where it meets energy and environmental needs and where it meets social needs.

This is not to say that conflict always can, or indeed should, be avoided. Some developers will continue to try to demolish worthy old structures, to commit new atrocities on the cityscape. If they can be reached early and encouraged to take an alternative development route, that is clearly preferable. But if they insist, they have to know that they will meet a public consciousness—perhaps in the form of a historical review process or zoning obstacles. Offered an olive branch, with the gloved fist held initially in reserve, devel-

Renaissance Center (1976-77, John Portman), a massive urban renewal project helping to change the face of downtown Detroit. (Photo: U.S. Department of Housing and Urban Development)

opers may be much more willing to deal with environmentalists and preservationists than in the past. They know their old suburban growth patterns are facing mounting practical obstacles, including environmental laws. They know also that the viable market is increasingly where preservationists are—in the established city or town.

There is a growth council in Santa Clara County, Calif., that includes developers, environmentalists, construction firms, unions and oftentimes minorities who testify at zoning hearings for approval of denser housing in established neighborhoods. Suddenly there is a countervoice to the selfish neighborhood that all so frequently says, Just leave us alone. We don't care about the needs of the rest of this city or metropolitan area.

Gentrification and displacement of the poor have become important issues in our cities. Preservationists and their allies can also try to help with solutions in these areas—such as zoning shifts, tax code changes or individual real estate deals—that protect the poor and elderly who want to stay, even as middle-class reset-

Main Street, Hot Springs, S.D. Noted for its sandstone buildings along the Fall River in the southern Black Hills, this main street was one of three in the pilot National Trust Main Street Project. (Photo: National Trust collection)

tlement of urban neighborhoods proceeds. Certainly the building of new townhouses or garden apartments on empty city lots helps the poor, because it makes their displacement from older buildings less likely. That kind of new housing matches the needs of the baby boom generation—especially because it often can be built at reasonable cost, with small square footages and on relatively inexpensive land. This reduces the outward pressure on farmlands. It saves public funds that would otherwise have to go into new infrastructure. And it saves energy on every count from individual household use to a reduction in commuting.

It may seem unnatural for preservationists to make common cause with developers. Jay Brodie, the Baltimore housing commissioner, has addressed this possibility by saying that the lamb and the lion may lie down together—but the lamb won't sleep much. Yet on issues such as the newer and denser housing that can be built in established urban areas, there is a great need for all—preservationist, city government, developer, architectural review commission—to work cooperatively toward high design standards and a good relationship between the new housing and the cityscape it enters. Around this country, there are increasing examples of handsome dense and mixed-use development but also too many examples of dull, characterless housing. Good design need not necessarily be gold-plated; the problem is to see that imagination and a commitment to quality are involved at the front end. Compactness, economy, harmonious and quality design can be achieved simultaneously, but only in an environment where the right skills are applied, the necessary political coalitions in place.

The same, I might add, is as true in small cities or towns as in big ones. The issues, indeed, are remarkably similar in all manner of places. Characterless sprawl development between towns, what some have called a linear suburban strip development, is as gluttonous in energy use, as harmful to the identity, the tax base, the vitality of the towns as is senseless sprawl out beyond city borders. Sparked by the National Trust, an important national effort is now under way to halp small towns and cities maintain the viability of their Main Streets.

Recently, National Trust President Michael L. Ainslie observed that preservation in the future must rely much more on developing a local constituency, on political connections such as getting preservation-minded people appointed to city planning offices or as aides to mayors. To save valued old buildings, he would like to see less emphasis on using the legal procedures in the National Historic Preservation Act of 1966 and the National Environmental Protection Act of 1969—because, among other things, there is now so much local resentment of federal interference. The legal safeguards are all too often after the fact, said Ainslie. It is better for preservationists to get themselves involved early in negotiations over new development and, once having participated and exerted their utmost, to live with the outcome. There was no question, he said, that the preservation community had on occasion used guerilla tactics to subvert projects it did not trust. Fundamentally, he added, preservationists "are in the real estate business—even if we're specialized in the old buildings department."

Needed: A Conservation Ethic

Is there a unifying principle behind all these ideas? I believe that there is. What our cities and towns need after the booming 1950s, the soaring 1960s and the misfired 1970s is a thorough-going conservation ethic. This does not mean that we stop or discourage growth. It means that we foster growth in a way that conserves energy, as energy in fact becomes the scarcest of all commodities. That we conserve our agricultural lands against unnecessary urban encroachment, both for the sake of our environment and because agricultural production is one of America's few strong points in the world economy today. That we work to conserve the buildings and texture of our cities and towns, because they reflect our culture and can become much stronger points of attraction for people in a world of necessarily more compact settlement. That we work to add on to our settlements—in inner-city development, in towns, in neighborhoods—in a way that matches the social needs of the times as well as excellence of design. That we conserve human resources, assisting less fortunate people to share in the exciting urban revitalization of this decade and giving them a helping hand toward self-sufficiency.

In all of this, the preservation community of our country should look to phenomenally broadened horizons: to the building of broad and complex coalitions, to the shaping of skills in every area from design review to insulation techniques to real estate marketing. This is a tall order, doubtless several leagues ahead of what the founders of the historic preservation movement would ever have dreamed necessary, much less possible. But given the exigencies of the times, it may be the only feasible course.

1. Saving Energy in Old Buildings

Introduction

Ann Webster Smith

"If primitive man could discover how to transform grain into bread and reeds growing by the river into baskets, if his successors could invent the wheel, harness the insubstantial air to turn a two-ton millstone, transform sheep's wool, flax and worms' cocoons into fabric, we, I imagine, will find a way to manage the energy problem," historian Barbara Tuchman commented during a 1980 National Endowment for the Humanities lecture, "Mankind's Better Moments."

These are cheering words, unless one stops to consider how long it took to invent the wheel in the first place, and how much time we spend reinventing it. One of the surprises of the current energy crisis is the fact that to millions of Americans, it does come as a surprise. This is the future shock for which so many Americans were totally unprepared. The notion that the disposable economy was one day going to disappear, much less that anything other than a disposable economy would be forced on us, was something with which we were not prepared to cope.

The environmental movement, the legislation that grew out of it and the great concern for the future of planet Earth should have given ample notice that all resources, not just air and water and land, were finite. For many, it is a painful process to learn that every form of energy can suffer the same process of disappearing, whether it does this as a result of individual human wastefulness or because of the vagaries of international politics.

Even though some Americans think that today's energy crisis is something that happened suddenly, similar disruptions have affected other societies, which have had to find ways to deal with them, too. In ancient Greece, forests near cities were destroyed in the search for wood to fuel cooking and heating fires and to build dwellings and the ships on which the economy of the society depended. So laws were passed to protect the firewood.

Ancient Rome suffered a similar problem. During the first century B.C., wood for use in the city had to be transported more than 1,000 miles. Then, 2,100 years ago, that distance probably was effectively farther than the distance from the United States to the Persian Gulf. During the 19th century, Californians found that the cost of heating water with natural gas was prohibitive, and artificial gas, which was infinitely more expensive than it is today, was introduced. Out of this particular energy crisis developed the first metal water tank, painted black and sited to gain heat from the sun.

We now have proof that the buildings with the poorest energy efficiency are those that were built between 1941 and 1970 and that old houses often use less heat and less energy for heating and cooling than do new houses. The reasons seem rather simple: Most old buildings have a ratio of 20 percent or less between glass and walls, providing only adequate light and ventilation; they use interior and exterior shutters, blinds, curtains, draperies and awnings to minimize heat gain or loss and they use such architectural features as balconies, porches and overhangs and such landscape features as shade trees and plantings around foundations to minimize heat gain.

Until recently, too many Americans had never focused on the notion that a building can work for them. But sometimes our cooperation is required. Not too many years ago, in my grandfather's house

Opposite: Albany, N.Y., historic row houses undergoing renovation on the author's street. (Photo: Louise Merritt, Historic Albany Foundation)

Columbia Congregational Church (1832), Columbia, Conn. Many old buildings were designed with shutters to reduce heat gain and loss. (Photo: © Philip Trager 1977)

in the South, the shades were always drawn in the mornings in the summertime, and both curtains and draperies were drawn at night. This was not just for privacy; it simply made sense for heating and cooling.

Several years ago when I moved to Paris, reality dawned on me during my first October there one day when I was shopping in a fashionable department store. I saw a wide range of ladies' woolen lingerie, some in high-style colors. Because I was working in a 17th-century building, I soon realized that the woolies of whatever style and color and the layered look, as it was called in the United States, had their basis in practicality. There is more than one way to conserve energy in old buildings.

In my own 1871 house in Albany, winter ended for me in effect when the adjoining row house owner completed his rehabilitation and turned on the heat. My house got warmer very quickly, and suddenly I needed less heat. Energy conservation is easier with a neighbor.

When I was growing up in Florida, I lived in a house with solar hot-water heaters. They worked well until the utility company came along and said that there was plenty of electric power and that there was no reason not to use it—in spite of the fact that the solar hot-water heat was free.

Today, solar heat is touted as a new technology, but in Greece and Rome, conservation and solar technologies were used in the same ways that they are used today and for many of the same reasons—because the alternatives are expensive and difficult to come by. So much for concern about reinventing the wheel!

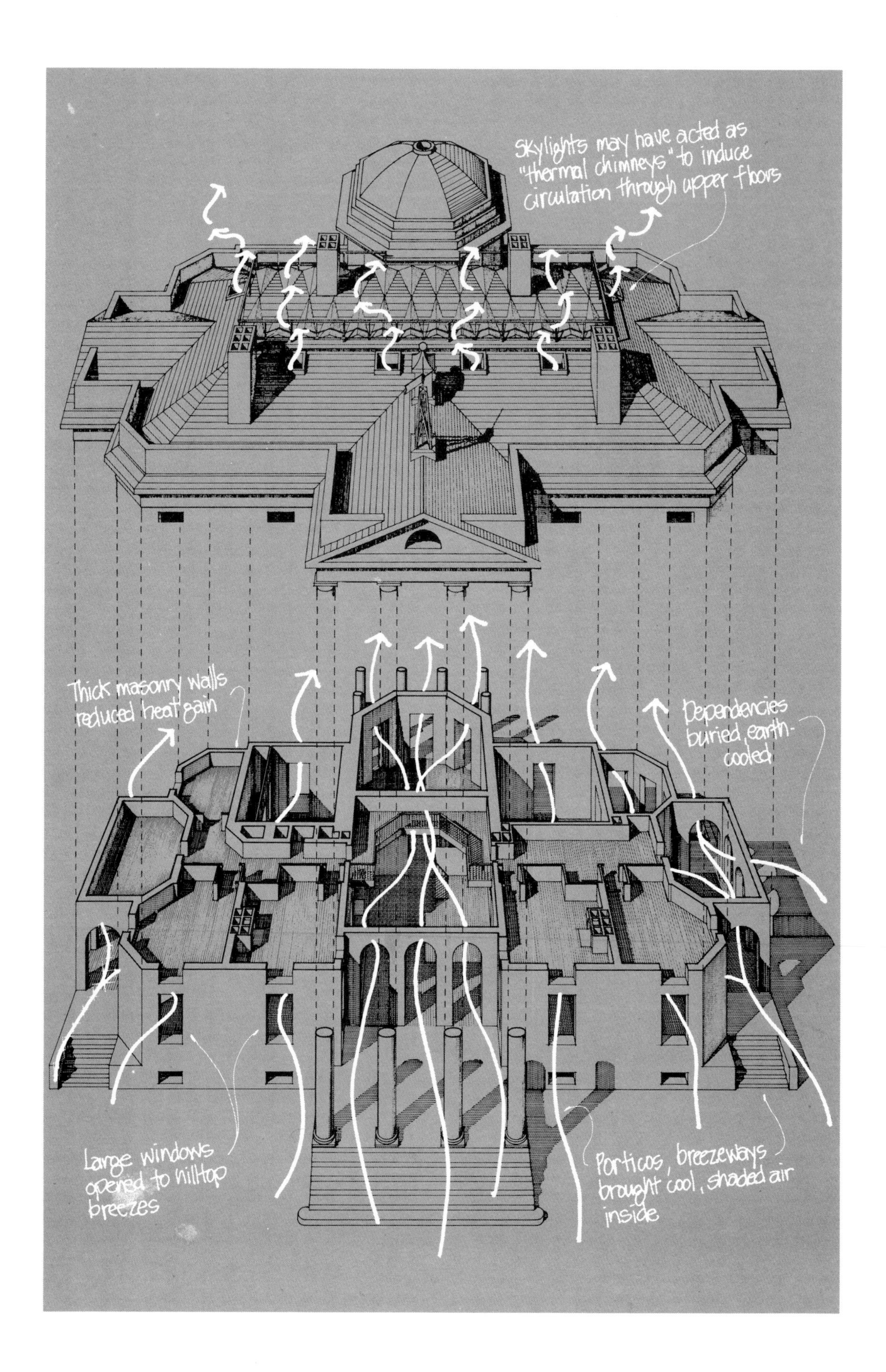
Skylights may have acted as "thermal chimneys" to induce circulation through upper floors
Thick masonry walls reduced heat gain
Dependencies buried, earth-cooled
Large windows opened to hilltop breezes
Porticos, breezeways brought cool, shaded air inside

Inherent Energy-Saving Features of Old Buildings

Theodore Anton Sande

Several observations on the inherent energy-saving features of old buildings should be made at the outset.

First, not every old building is energy conservative. The purpose for which a building was originally intended must be considered, and that purpose may have been far removed from the idea of saving energy. For example, a medieval French castle was shaped primarily by concerns for defense; an Italian Renaissance palazzo by the desire to express political power, albeit by graceful and artistic means; and a late 19th-century American "cottage" at Newport, R.I., by the ambition to demonstrate how much wealth its owner possessed and how lavishly it could be spent—in Thorstein Veblen's terms, "conspicuous consumption." This does not mean that the resulting buildings were always wasteful in themselves, rather that they stand for values different from those associated with the economical use of energy.

Second, when the issue of energy conservation in old buildings is addressed, there seems to be an assumption that energy conservation is considered in relation to something else—and that something else is habitability. There must be an equation between how much energy is used by a building and the extent to which this use of energy enables a building to serve its intended purpose efficiently. And ideas about habitability are most readily apparent in our dwellings.

Last, for much of the world the concept of habitability has shifted markedly over the past 200 years, since the emergence of industrialism in the early 18th century. This shift started slowly, gradually gaining momentum until it got into high gear by the last quarter of the 19th centruy. It has continued virtually unrestrained until relatively recently, when fuel suppliers in the Middle East became uncooperative in providing inexpensive petroleum.

The shift discussed here is one toward the increasingly higher levels of comfort that have resulted from a modern, technologically based society. For example, an 18th-century English manor house was certainly more elegant than a farmer's cottage of the day, but one was little different from the other with respect to ambient interior temperature on a cold, gray winter's afternoon. If anything, the farmer's cottage was probably the more cozy and energy conservative of the two. By the mid-19th century, wood and coal-fired central heating systems were replacing the hearth; these were followed by municipal gas heating and lighting, oil burners, electrical lighting and power circuitry and, finally, centralized air conditioning's complete environmental control of air temperature, humidity and flow. As a result, a relatively modest suburban house of the late 20th century possesses a range of physical comfort and environmental control that would have been unattainable for the lord of the 18th-century English manor house.

Three questions may be asked to help clarify our understanding of the inherent energy-saving features of old buildings:

1. To what extent did our ancestors think about energy conservation in their houses?
2. How is their concern evident in buildings that have survived?
3. What energy conservation lessons can be learned from these buildings?

Opposite: The natural air-conditioning system designed into Monticello (1770-1809, Thomas Jefferson), Charlottesville, Va. (Drawing: From Research and Design. *Courtesy Thomas Jefferson Memorial Foundation)*

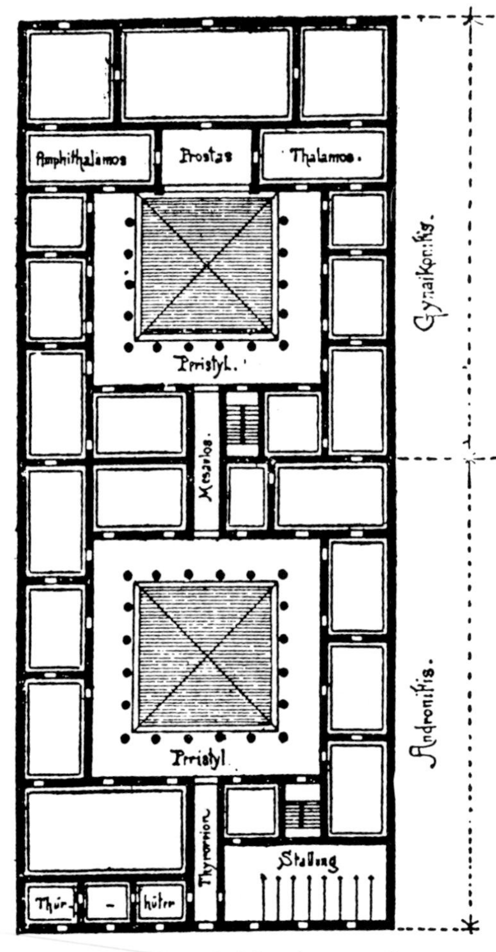

Plan of a Greek house by Vitruvius, according to Becker. (Drawing: From The Ten Books on Architecture*)*

Question 1: Early Design Concepts

The answer to the question of how much our ancestors thought about conserving energy in their houses depends on the inferences drawn from early written accounts of architectural practice. The term "energy conservation" is not apt to be found in old texts, but it is safe to assume that these writings, where they refer to principles of design, imply *efficient* design; furthermore, they are always careful to relate a building to its natural environment. So it can reasonably be concluded that these earlier houses were meant to be relatively self-sufficient and not to require a great expenditure of energy for their maintenance or comfortable occupancy.

In the Roman world of the first century B.C., an average practitioner of his time, the architect and engineer Marcus Vitruvius Pollio, devoted chapter 1, book 6, of his *The Ten Books on Architecture* (reprinted by Dover Publications, 1960) to the topic "On Climate as Determining the Style of the House." He states as the very first sentence of this chapter, "If our designs for private houses are to be correct, we must at the outset take note of the countries and climates in which they are built."

Chapter 2 discusses modifications to suit the site and chapter 3, the proportions of the principal rooms of a house. But it is chapter 4 that is of greatest interest, for here Vitruvius sets out his rules for the proper exposures of the different rooms:

> 1. . . . Winter dining rooms and bathrooms should have a southwestern exposure, for the reason that they need the evening light, and also because the setting sun, facing them in all its splendour but with abated heat, lends a gentler warmth to that quarter in the evening. Bedrooms and libraries ought to have an eastern exposure, because their purposes require the morning light, and also

Øygarden Farm (late 18th century), Lillehammer, Norway, built to use the earth's heat and coolness. (Photo: Lee Nelson)

Lafayette's headquarters (c. 1763), Chadd's Ford, Pa., near the Brandywine Battlefield. (Photo: Ned Goode, Historic American Buildings Survey)

because books in such libraries will not decay. . . .

2. Dining rooms for Spring and Autumn to the east; for when the windows face that quarter, the sun, as he goes on his career from over against them to the west, leaves such rooms at the proper temperature at the time when it is customary to use them. Summer dining rooms to the north, because that quarter is not, like the others, burning with heat during the solstice, for the reason that it is unexposed to the sun's course, and hence it always keeps cool, and makes the use of the rooms both healthy and agreeable. . . .

Similarly, one could quote others who repeated or built on these ideas down through the years. Leon Battista Alberti, the great Italian architect and theorist of the early Renaissance, in 1450 wrote about the perfect country house emulating Vitruvius and adding commentary about fireplaces, stoves and chimneys. In the United States in the 19th century, a number of persons wrote about architectural design from the points of view of health and comfort. One of the most well known was Andrew Jackson Downing, whose *The Architecture of Country Houses* appeared in 1850.

The point of all this is that, up to the mid-19th century, those who thought seriously about residential design did so within the context of long-established precedents that took into account environmental factors in planning houses and that, to the extent that a house was compatible with its site and climate, it probably used energy efficiently.

Question 2: Evidence of Conservation

On the question of available evidence of energy conservation in surviving buildings, it is helpful to examine old buildings under four subheadings: siting, exterior

Stratford Hall (c. 1725), Westmoreland County, Va., with its raised main floor. (Photos: Jack E. Boucher, Historic American Buildings Survey)

Folding wood-paneled window shutters in a ground-floor room at Stratford Hall.

Kitchen dependency of Stratford Hall, located away from the main house.

Draperies, shutters and canopied bed used for controlling heat gain and loss in the mother's bedroom on the main floor of Stratford Hall.

Monticello (1770-1809, Thomas Jefferson), Charlottesville, Va., sited and designed to take advantage of natural energy conservation factors. (Photo: John J.G. Blumenson, National Trust)

configuration, interior configuration and sun control.

Siting. A late 18th-century Norwegian farm complex at Lillehammer, with wood-sided, turf-roofed outbuildings, shows varying relationships between the structures and the ground surface and illustrates the ways in which structures have been placed to use the earth's heat or coolness for storing perishable foods and other items. The exposed wall surfaces were reduced to mitigate the effect of severe winter winds. At Lafayette's residence and headquarters, Chadd's Ford, Pa., 18th-century masonry farm buildings are clustered together on the gently rolling terrain—providing mutual protection against the elements and sheltering a southern-exposed, sun-warmed courtyard beyond. The main house at Stratford Hall (c. 1725) in Westmoreland County, Va., the birthplace of Robert E. Lee, illustrates an arrangement in which high-ceilinged ceremonial or entertainment rooms are placed above a ground floor of lower-ceilinged bed chambers and utilitarian rooms. Thus, summer could be spent in the relative coolness of the ground floor and winter upstairs on the first floor. The house is also well sited to achieve cross ventilation by taking advantage of prevailing breezes from the Potomac River to the north. The kitchen, with its ever-present threat of fire and attendant nuisances of heat and odors, is located in a separate, flanking dependency.

Thomas Jefferson gave considerable thought to the design of his house, Monticello, between 1770 and 1809. He placed it on the eastern side of a leveled plateau crowning a mountain, its main facades facing east and west and its long, low wings sunk into the terrain and projecting out and around to the north and south,

Richard Jackson House (1664), Portsmouth, N.H., showing its central chimney. (Photo: Richard Cheek, Society for the Preservation of New England Antiquities)

terminating in two-story end pavilions. The low wings contained offices, horse stalls, a carriage house, a washroom, wine and beer rooms, servants' quarters and a smokehouse. The kitchen also was at the lower elevation, but in a separate structure off the southeast corner. A recent study suggests that Jefferson experimented with air circulation for increased comfort in the main house itself (see "Passive Cooling," *Research and Design*, fall 1979).

Exterior configuration. Two 17th-century colonial American houses—Bacon's Castle (1665), Surrey County, Va.; and the Richard Jackson House (1664), Portsmouth, N.H.—illustrate the main features of the second category. Both are based closely on English medieval provincial precedents from the respective regions of their colonial settlers. "Exterior configuration" means the arrangement or design

Kitchen hearth in the Richard Jackson House. (Photo: J. David Bohl, Society for the Preservation of New England Antiquities)

Bacon's Castle (1665), Surrey County, Va., showing chimneys placed in the gabled ends. (Photo: Jack E. Boucher, Historic American Buildings Survey)

Jabez Howland House (1667), Plymouth, Mass., designed around the large fireplace to capture available heat. (Photo: Pilgrim John Howland Society)

of principal architectural features, such as chimneys, projections from or recessions in a facade, roof and wall materials and the size and arrangement of windows and doors.

It would be misleading to suggest that the differences between these two roughly contemporary houses are attributable entirely to responses to the different climates in which they were built, but there is enough of that consideration present to make a useful comparison. For example, the main portion of the Richard Jackson House, in a northern climate, has a central chimney so that heat radiates from the fireplaces, warming the surrounding rooms. These rooms and the clapboard-sheathed sides and steeply pitched, wood-shingled roof are meant to capture and retain as much heat as possible in the bitter, damp coastal weather of New England.

At Bacon's Castle, located in a southern, more moderate climate, the chimneys are placed at the gabled ends. Heat is important here as well, but not to the extent required for New Hampshire. Even more important is the need to dissipate heat in the warmer months of the year, and end chimneys give up their heat more quickly than do those in the center of a dwelling. In the 17th century, fireplaces were used all year for cooking and not just to provide warmth. (As at Stratford, more affluent owners later removed the kitchen entirely to a separate outbuilding.) The thick masonry walls, parged or plastered on the inner faces, were intended to give better insulation against the hot sun and to keep the interior cooler in summer. Windows tended to be smaller and fewer in the North than the South, although size depended as much on the expensiveness and scarcity of glass as it did on any climatic considerations. Doors, in both regions, usually were placed in the center of a facade, opening into a small hallway or "porch," as it was called in the 17th century, that kept cold air from blasting into the major hall or parlor. The doorway pediment in the center pavilion of Bacon's

Hanbury Hall (early 18th century), near Droitwich, Hereford and Worcester, England. (Photo: National Trust)

Castle is still visible, although a ground-floor window now is located there.

Interior configuration. The interior of the Jabez Howland House (1667), Plymouth, Mass., a typical 17th-century northern house, provides an example of the third category. The fireplace is huge, both to generate as much heat as possible by this generally inefficient means and to allow sufficient space for cooking. The ceiling was kept low to contain heat, and a high-backed bench, a "settle," was placed close to the fire, its solid, tall back designed to contain and reflect the warmth from the fire around the hearth, where the inhabitants spent much of their winter life.

In sharp contrast is the interior of a main-floor bedroom at Stratford Hall, built 60 years later in a much warmer climate. The windows have draperies, not just because they are decorative, but because they may be used to control drafts, keep out sun or keep in heat. (Wood-slatted venetian blinds also were used for this purpose in the 18th century.) At the sides, or jambs, of the windows, folding wood-paneled shutters tuck into the wall. This was the most effective way to control heat loss or gain at the window, and it is sensible for the shutters to be located on the inside where they are accessible, rather than on the outside. The furniture as well is important at Stratford, where a canopied bed with side curtains provided another means during cold weather for keeping in warmth—in this instance, body heat.

Sun control. The fourth category involves consideration of porches, roof overhangs and general orientation of a building. Hanbury Hall, Hereford and Worcester, is an early 18th-century English country house. Its facades, particularly the southern one, are distinct from those of its predecessors in that the ancient requirements of defense, which had meant either no windows or only narrow slits in a wall, had diminished to the point that one could, with reasonable assurance of retaining occupancy, put up generous amounts of glass and enjoy the view.

Dunleith (c. 1855), Natchez, Miss., with its broad two-story porch. (Photo: Mississippi Department of Archives and History)

Interior of Augustus Saint-Gaudens's Little Studio (converted 1900), Cornish, N.H., c. 1926. (Photo: Saint-Gaudens National Historic Site)

Porch of Lyndhurst (1838, 1864-65, A.J. Davis), Tarrytown, N.Y., in a drawing by the architect. (Drawing: National Trust collection)

However, the facades are still planar, with nothing projecting from them. It really was not until the first half of the 19th century and the Gothic Revival in Europe and America that the veranda began to be introduced as a transitional device between the interior environment of a house and the natural landscape beyond. Although the porch had deep philosophical links with the Gothic Revival and with the beliefs of transcendentalism, its comfortable virtues were recognized quickly, as evidenced by the porch that still surrounds three sides of the southern portion of Lyndhurst (1838, 1864-65, A.J. Davis) in Tarrytown, N.Y. It was soon discovered that other architectural styles from the past either used porches in one fashion or another or could be modified easily to accommodate them. The South, for example, adapted the Classical Revival styles to suit its climatic needs. Thus, at Dunleith (c. 1855), Natchez, Miss., the Classical Revival was expressed in a broad, umbrella-like two-story porch totally encompassing the living spaces within. People still like porches, as reporter Dick Dabney lamented in the *Washington Post* (May 6, 1980):

> We were discussing front porches generally, and there too we were of one mind in that we loved them and believed that they had been killed off most likely by machines: television, which had destroyed conversation, and air conditioning, which let people stay inside when they had no business there; and even automobiles, which had fixed it so there wasn't anybody strolling by your front porch anyway who could come up and pass the time of day.

Augustus Saint-Gaudens's studio at Aspet, built in central New Hampshire just east of the Connecticut River in the late 19th century, also shows the application of sun-control principles, drawing light into the studio high at the upper level, yet suppressing it at the ground floor by means of a covered walk and grape arbor.

Question 3: Lessons from Old Buildings

What lessons about energy conservation can be learned from old buildings? We can draw at least one conclusion: The extent to which a building's design takes into account its site and climate determines how efficiently it will use whatever energy is employed for making it habitable.

Furthermore, nothing in the available evidence on old buildings suggests that it costs one cent more to design a building so that it properly fits its site and climate. Taking advantage of the earth's heat and coolness, facing a main facade to capture the sun's rays, constructing a porch to shade a facade, placing a heating system to be most efficient or sizing windows to let in light yet minimize heat loss—all of these considerations are matters of judgment and planning and do not require great expenditure for their attainment.

In short, the first and fundamental principle we must learn from old buildings constructed before the widespread use of modern heating, ventilating and air-conditioning systems is that, in general, they were built with consideration for site, environment and climate.

Perhaps Frank Lloyd Wright, who matured during the period in which modern environmental-control technologies were developing but who also read Downing, William Morris and John Ruskin, expressed this best when he wrote in *An American Architecture* (Horizon Press, 1955): "Man takes a positive hand in creation whenever he puts a building upon the earth beneath the sun. If he has birthright at all, it must consist in this: that he, too, is no less a feature of the landscape than the rocks, trees, bears or bees of that nature to which he owes his being."

Vitruvius would have understood this sentiment, too.

Making Buildings Work as They Were Intended

Baird M. Smith, AIA

The remarks in this paper address energy conservation problems in a variety of building types—the typical wood-frame individual building, the masonry structure, large-scale buildings and groups of buildings, both residential and commercial in nature. Although the material refers to "historic buildings," much of what is said applies equally to any old or existing building, of which there are many thousands throughout the country.

Building Characteristics

In any consideration of energy conservation, old buildings often are viewed simply as big, old and drafty. It is commonly assumed that they are probably expensive to heat and that major retrofitting actions will be required to conserve energy. Two points contradict this common perception.

First, there is evidence that many old buildings use far less energy than new ones. A detailed analysis of New York City office buildings, selected from a sample of about 1,000, revealed that the oldest buildings use much less energy than the buildings constructed after World War II. There is a reason for this. The old buildings probably were colder in the winter, and the people working in them may have had to wear sweaters. They probably were warmer in the summer, forcing men to loosen their ties. The mechanical systems that had developed in the buildings were a mishmash, often with a combination of interior heating equipment and window air-conditioner units. As a result, the rooms often were individually controlled. The occupants could turn the thermostats up or down; they could even open the windows. This, of course, is almost radical to office building design of the 1960s and 1970s.

Opposite: Interior court of the Pension Building (1882-85, Montgomery C. Meigs), Washington, D.C., designed for natural light and ventilation. (Photo: Walter Smalling, Jr., U.S. Department of the Interior)

Second, these buildings can be improved but the inherent qualities of a building—those characteristics discussed by Theodore Sande in the preceding paper—should be studied before any retrofitting action is begun. This need is well illustrated by the Pension Building (1882-85, Montgomery C. Meigs) in Washington, D.C., a unique structure and certainly not one typical throughout the country. The design is high style, but many people do not realize that the building is technologically very advanced. The construction is masonry, and there is a low ratio of glass to wall (approximately 20 percent).

The windows—individually large, about eight feet high—were sized to admit a proper amount of light into each space. As most of the workrooms are illuminated by sunlight, electric lights actually are supplementary. The windows provide ventilation when open and, when closed, fresh air could enter through air vents below the windows. The air that enters the building through the offices flows out through the clerestories that are above the interior court. When the building was first occupied, the flow of air through the offices was so great that workers could not keep papers on their desks, and the design had to be modified. Originally, the window specifications called for double glazing, but the high cost caused this to be eliminated.

However grand this building is, it should be reemphasized that it is a technologically advanced building and that it is unique. But the point remains: If the attributes of historic buildings are considered and allowed to function as they

Energy Consumption in New York City Office Buildings

Construction Date	Number of Buildings	Percentage of Buildings	Percentage of Area	Energy Consumption Range (MBtu/sq. ft.)	Average Consumption (MBtu/sq. ft.)
Before 1900	3	6.8	1.1	83-115	95
1901-19	8	18.2	12.8	76-135	105
1920-40	18	40.9	28.3	68-223	109
1941-62	12	27.3	36.2	66-198	126
1962-70	3	6.8	21.6	78-163	115

Based on in-depth analysis of 44 sample buildings selected from 1,000 contacted and 436 studied further. Energy consumption was measured by evaluating expenses for utilities (electricity, fuel oil, coal, etc.) from 1970 to 1975.

Source: *Energy Conservation in Existing Office Buildings*, by Syska and Hennessy and Tishman Research Corporation for U.S. Energy Research and Development Administration, 1977.

were intended, a great deal of energy may be saved without any retrofitting.

Some General Recommendations

All the windows should be functioning correctly. Cupolas and monitor windows were intended to allow air to exit in the summer; these also should be operative. Vents—roof vents, attic vents or others—should be open and not blocked. Shutters should be used if possible; they were intended to be opened and closed, as necessary.

Awnings were popular during the 1890s and through the early 20th century. They are appropriate for almost any historic building, and there probably is evidence for awnings on most old structures. It should be emphasized that any exterior treatment that prevents the sun from entering a window, such as an awning or a shutter, is seven times more effective at reducing heat gain than is any device on the inside of a window. Thus, awnings, shutters or shade screens are by far the most effective means to reduce heat gain.

Landscape features should not be forgotten. In the summer, deciduous trees are extremely beneficial in helping to reduce heat gain. Pine trees and other evergreens can be used as windbreaks or to reduce the severity of winter storms. Any of these features should be considered and developed.

Actual building use—how a building or home is operated—can help to conserve energy. For example, the thermostat setting for the furnace and the number of hours occupants are in certain rooms will greatly affect the quantities of energy used. Measures to identify how and when individual rooms are used, and steps to control the heat supplied to these rooms, will reduce overall energy usage often by as much as 30 percent. The point is to heat or cool only those rooms actually used, and only for the periods during which they are in use.

By this stage of analyzing energy conservation measures, an awareness both of the building's attributes and of the means to make the building work as it was intended to work should be obvious. In addition, operational controls can be instituted after an energy audit.

New York City high rises in the vicinity of Rockefeller Center (1940, Reinhard and Hofmeister; Corbett, Harrison and MacMurray; Hood, Godley and Foulihoux). (Photo: Courtesy Rockefeller Center, Inc.)

Bottom: Old view of the Pension Building (1882-85, Montgomery C. Meigs), Washington, D.C., when awnings were used to reduce solar heat gain. (Photo: Veterans Administration collection, National Archives)

South facade of the Pension Building, showing the low glass-to-wall ratio, large windows and clerestories designed for natural air circulation. (Photo: Walter Smalling, Jr., U.S. Department of the Interior)

Detail of a Pension Building window with fresh-air vents used for circulation when the windows are closed. (Photo: Baird M. Smith)

Windows and roof vents such as those on this Denver house can be used as they were intended for controlling air circulation and heat gains and losses. (Photo: Carleton Knight III, National Trust)

Energy Audits

As retrofitting is considered, careful evaluation of the building should be undertaken. Such a survey should begin with a determination of the U-value or R-value for the walls, windows, ceilings and roofs. In historic buildings, this can be difficult because the exact construction of the wall, the materials and composition, often are unknown. Only physical probing can determine whether a wall is a cavity wall, if it is brick or hollow clay tile, if insulation is present, etc. A specific process or an audit is required to determine the thermal characteristics of these materials. Sometimes, probing will reveal unfamiliar insulation materials; for example, in one unusual case, straw in a light plaster base was used in a ceiling. Such things as newspaper, sawdust or wood chips and cinders often were integrated into walls to increase their thermal resistance.

Compounding such evaluative difficulties may be a phenomenon referred to as thermal mass. In a masonry building, heat is stored in the walls during the day and released slowly during the night, complicating the calculation for determining the thermal resistance of materials. Many designers, engineers and architects are not familiar with the corrective factors for thermal mass and often use incorrectly the tools that are available to them.

In considering other techniques that can be used, it should be emphasized that one advantage of historic or existing buildings is the fact that they are in place. One can look at them, walk through them and measure some of their qualities and characteristics. For example, infrared

Fisher Brothers Store (1856) and Bella Union Saloon (1856), Jacksonville, Ore. A variety of awning types are appropriate conservation measures in contemporary usage. (Photo: Jack E. Boucher, Historic American Buildings Survey)

Bottom: Nesmith-Greeley Building (1888), San Diego. During the 19th century awnings were popular energy conservation features on commercial buildings as well as houses. (Photo: San Diego Title Insurance and Trust Company collection)

Oakleigh Garden Historic District, Mobile, Ala., landscaped with deciduous trees that provide energy conservation. (Photo: Carleton Knight III, National Trust)

Ceilings such as this are often lowered in the hope that less energy will be needed to heat the smaller space, but the result is dramatic and inappropriate change to a historic building. (Photo: Baird M. Smith)

photography often is used to identify hot spots, which appear in the photograph as bright white. In one such survey, the windows, obviously a great source of heat loss, showed up bright white, but bright white areas also appeared below each window. It was discovered that radiators were below each window, a typical configuration. Construction of a simple radiator cover that reflects heat back into the room vastly increased the efficiency of the heat system without any extensive retrofit.

Conservation and Building Codes

Various energy conservation standards or codes come into play in the consideration of any substantial rehabilitation. The application of these modern standards to old buildings produces sometimes disturbing implications. Some are highly prescriptive. For example, they identify a specific U-value or heat transmission standard that should be attained in a wall element, a window or a roof. Clearly, this may be difficult to achieve when one is dealing with an existing building, because the flexibility to alter the materials does not exist.

Most of the standards were written for application to new construction. Increasingly, the codes developed for new construction are applied inappropriately or incorrectly to existing buildings. Energy standards that are specifically directed toward existing buildings, recognizing their historical character, are needed.

Inappropriate actions often result from the application of inappropriate standards. In a room with a high ceiling, for instance, the ceiling may be dropped in the belief that this is an effective conservation move. Another common measure is the removal of a building's windows. Windows are the weakest thermal element of a building; thus, the first treatment in any substantial rehabilitation often is the removal of existing windows and the installation of a double-glazed system, often inappropriate to historical character. Such changes also are expensive, and it has been shown that retention and repair of original windows plus the installation of storm windows provides a thermally superior combination at much less cost.

What we must attempt to achieve is a recognition of the historical character of buildings, in particular those characteristics that are thermal attributes. We should be able to retain the awnings, the windows, the historical features and hence help attain one of our national goals—the conservation of cultural resources. We also should be able to improve the thermal characteristics of a building through operational controls or retrofitting and, therefore, attain a second national goal—the conservation of energy.

An Old-House Conservation Strategy

Nathaniel Palmer Neblett, AIA

Now that the nation's energy needs and resources demand a conservation ethic, the most successful approach to energy management in the historic building is one designed to maximize the advantages, minimize the faults and add or alter only what is necessary to achieve acceptable results. A realistic assessment of existing conditions is the first step in any conservation plan.

Developing the Strategy

Orientation. Start by considering the building's orientation. Keep in mind that the sun rises in the east, sets in the west and is higher in the sky and of longer duration in summer than in winter. Naturally, the southern exposure will be the sunniest, and any action undertaken either to intensify or to negate solar effects will be concentrated on the southern side of the house. If the layout of the house affords flexibility, true conservationists will arrange their mode of living to follow the sun in its daily course. This adjustment to nature will allow a savings in energy both for heat and for light.

Roof. The function of the roof is to provide an overhead barrier impenetrable to all hostile elements. Over the years, a number of materials have been employed as roof surfaces; however, the important consideration from the energy conservation point of view is that whatever roof is installed is doing its job. Any moisture that finds its way into the thermal insulation lowers the ability of the insulation to resist heat transfer.

Opposite: Ivinson Mansion (1892, W.E. Ware), Laramie, Wyo., with its energy-conserving porch, dark roof, overhangs and working windows. (Photo: Jack E. Boucher, Historic American Buildings Survey)

The design of the roof may provide energy-saving opportunities also. Bold overhangs shield the windows below from the sun's rays in the hot summer but let the winter rays penetrate when the sun is lower in the sky. The overhang also tends to throw rainwater away from the structure, preventing its soaking into the wall and lowering the insulation value.

Color and texture influence the roof's reaction to solar effects. A dark, rough surface will absorb much more heat than will a light, smooth one. These factors should be kept in mind if a change in roof material is being considered.

Openings. Windows and doors penetrate the building shell to allow access for inhabitants and light and air. In addition to the principal closing device, these openings may be fitted with auxiliary devices such as awnings, blinds, shades or shutters that help keep out intruding elements. Assure that all such devices are in proper operating order and that they are used systematically for the functions intended. Surprising savings in energy consumption can result from the regular and proper use of these auxiliary closures. Windows located on the temperate side—the east and the south—admit the warm, cheery sunlight and make the interior more inviting.

Porches. Porches are designed to meld the advantages of interior and exterior. They can be real energy savers when properly and regularly used. Time when the family gathers on the porch to enjoy the delights of cooling breezes is time when the inside cooling apparatus may be turned down or off. Porches also act as a buffer to shield the main part of the house from the harsher aspects of the weather. The season of a porch's utility can be extended by the judicious use of jalousies or removable panels. A porch so enclosed may act as a sun pocket that is comfort-

Carpenter Gothic porch on a house in Boalsburg, Pa., a natural energy-saving architectural feature. (Photo: Alison Taggart. From Historic Buildings of Centre County, Pennsylvania. *Pennsylvania State University Press, 1980)*

able on clear winter days without auxiliary heat.

Landscape. Considerable energy conservation can result from a properly planned and managed landscape. Deciduous trees planted on the east, south and west sides of the house will provide screening from the sun's bright rays in the summer, yet allow the solar warmth to penetrate during the winter. A dense hedge of evergreens will provide a shield against winter winds that otherwise strike the building with full force. A garden wall adjacent to a patio will absorb and radiate the warmth of the winter sun to provide a delightful outdoor living space usable a major portion of the year. Plant material should be kept well away from the face of the building, and vines should be stripped from any building walls. In addition to the destructive effects on the historic fabric, this vegetation tends to keep the wall damp and thus lower its insulating value.

Analyzing Sources of Waste

Tests have been conducted to determine the greatest source of heat loss in old houses. If one takes a typical old frame house that has undergone no treatment for heat conservation and is two stories above a full basement, has double-hung windows with no weatherstripping and has no wall or attic insulation, the heat loss averages are as follows:

Loss at doors and windows: 50 percent
Loss through ceiling, attic, roof: 25 percent
Loss through walls: 18 percent
Loss through floor and basement: 7 percent

It becomes readily apparent from these test results that first conservation dollars are best spent attending to the heat loss at windows and doors and through the ceiling.

Shuttered and shadowed windows on a house in Carlisle, Pa. (Photo: A. Pierce Bounds)

Heat transfer. A basic knowledge of the physics of heat is necessary for devising the most efficient conservation plan. It must be remembered that heat is constantly flowing from masses of higher temperature to masses of lower temperature. Three methods of heat transmission have been defined:

1. Conduction
2. Convection
3. Radiation

Heat moving through a substance from molecule to molecule is heat flowing by conduction. The greater the difference in temperature, the greater the rate of flow.

Rising currents, set up by a hot object because heated air is less dense and tends to rise, carry heat through convection. A cast-iron radiator transmits heat to a room primarily through convection. Any material that will act as a barrier to the flowing currents will counteract heat transfer through convection.

The flow of heat through space between two bodies without heating the intervening space is radiation. The warmth felt on the face of a person sitting before an open fire is an example of heat transmitted through radiation. Shiny, reflective surfaces repel heat transfer by radiation, whereas dull, dark surfaces promote it. Light also is reflected by surfaces that are smooth and light colored. Energy may be saved by taking full advantage of this physical principle. Walls and ceilings painted white or a light tint will reflect the light, thereby achieving the desired level of illumination by using a light source of less intensity. Dark shades and pronounced textures absorb the light and give the room a gloomy appearance.

Illumination. Americans tend to over-illuminate inhabited spaces. The soft patina of older houses and furnishings is enhanced by the glow of low-intensity light sources. Certainly sufficient, well-directed light is necessary to prevent eye strain when reading or other close visual

Col. William Rhett House (1712), Charleston, S.C., with its partially enclosed upper porch, extending the time in which it can be used. (Photo: Louis I. Schwartz, Historic American Buildings Survey)

tasks are undertaken. More pleasant living spaces would result, nonetheless, from a lower level of general artificial illumination, and the energy savings could be significant. A 100-watt bulb consumes 67 percent more electric power than a 60-watt bulb. In each location where a 60-watt is substituted for a 100-watt bulb, savings will be effected.

Humidity. Although the human body feels more comfortable when the relative humidity is around 50 percent, frequently 10 percent or less is encountered inside heated homes. By increasing the relative humidity to 50 percent, the apparent temperature may be increased several degrees, thereby saving fuel. A properly operating humidifier on a forced hot air heating system is an effective device for raising the humidity, although a higher level might be achieved also with a self-contained humidifier or even with pans of water. A word of caution is in order here: The interior humidity should not be raised to the point that water condenses on cool surfaces. At this level, the moisture will begin to have detrimental effects on the structure.

Accentuate the Positive

Air circulation. All adjacent rooms in constant use should be open to each other to allow free circulation of conditioned air. This tends to eliminate hot and cold spots. A small electric fan can be used to set up currents that will bring the hot air from the ceiling down to the lower spaces where the heat is needed. The old type of circulating fans that are regaining popularity are used frequently for this purpose in old houses and on porches.

For summer comfort, air circulation is most important. One should consider the merits of an attic exhaust fan to expel the hot air of the daytime and replace it with the cooler evening air from outside. Frequent use of such a fan will obviate the need for an elaborate mechanical cooling system.

Temperature difference. As the difference between inside and outside conditions increases, a building is subjected to increasing stresses. Energy consumption jumps with every degree of difference. The conservationist's goal should be to lessen that difference as much as possible, yet maintain a healthful environment. One might try setting the thermostat at 60 or 65°F during the heating season and wearing warmer clothing inside. It is now generally recognized that a daytime setting of 65°F and a nighttime setting of 60°F will produce a temperature level

Bottom: Plan of Valley View (c. 1850), Cartersville, Ga., showing how landscaping can be used to help minimize heat gain and loss. (Drawing: From Garden History of Georgia, *Loraine M. Cooney, compiler. Courtesy Peachtree Garden Club, Atlanta)*

Landscaping: Energy-Conserving Locations

Landscape Elements	Cold Climate	Temperate Climate	Warm Climate
Ground cover or grass	negligible effect on all sides	on south	on east, west and south
Paving	on south	shaded if on south	shaded if on east, west and south
Shrubs against house wall	on east, west and north	on east, west and north	on all sides
Deciduous shade trees	negligible effect on all sides	on south and west	on east and west
Evergreen trees	on east, west and north	on east and west	on east and west
Windbreak (trees, bushes, fences)	on sides exposed to winter winds	on sides exposed to winter winds	undesirable on all sides
Windbreak used to funnel wind	undesirable on all sides	on sides exposed to summer winds	where cross ventilation is possible

Source: "Energy Conservation with Landscaping," by Richard Crenshaw. Federal Energy Administration with National Bureau of Standards.

Simple energy-saving steps such as weatherstripping, caulking, shading devices and using built-in energy conservation features will help reduce energy consumption in old buildings. (Drawing: From A Primer: Preservation for the Property Owner. *Preservation League of New York State, 1978)*

adequate for most human endeavors. The furniture, decorative objects and house plants also will respond favorably to the cooler environment.

Closures. Closures available in the house should be used to the fullest extent. Shades, curtains and blinds on the sunny side of the house should be opened to gain maximum solar effect. To minimize heat loss, others should be closed when the room is unoccupied. Any rooms that are unused over an extended period should be closed off from the rest of the house. The heat source in the rooms should be cut down or turned off entirely, depending on individual conditions. If the house has a vestibule, it can be used to full advantage by keeping all doors shut.

The important point to remember is that every layer of material, no matter how thin, between the conditioned space and the outside contributes to lessening the transfer of heat. Large pieces of case furniture, such as bookcases filled with books, placed against an outside wall will add insulation value.

Chimney flues should be closed when not in use. If left open all the time, the heat loss effect is about equal to that of an open window. Flues that are never used should be closed with a blanket of insulation to achieve an even greater degree of heat conservation. They should be checked periodically to assure that the damper is closed tightly. Dirty flues can sap energy dollars. Chimneys should be checked for cleanliness at least annually and soot deposits removed immediately in the interest of safety as well as energy conservation.

Well-maintained equipment. Equipment must be maintained in top operating order to secure maximum efficiency. All components of the heating or cooling plant should be thoroughly checked at the beginning of the season. The efficiency of the flame should be tested and burner nozzles changed when necessary. Working parts should be properly lubricated and chains and belts checked for wear. Filters must be cleaned or replaced at recommended intervals. The efficiency

Living room, Pope-Leighey House (1940, Frank Lloyd Wright), Mount Vernon, Va. Bookcases against the outside wall add insulation value. (Photo: Carleton Knight III, National Trust)

of cast-iron radiators may be reduced by 25 percent or more either by scale on the inside or thick paint build-up on the outside. One coat of paint—certainly no more than two—is all that should be allowed to accumulate. Heat-resistant paint with metallic pigments is satisfactory for use on radiators when they need to be repainted. Water-heater tanks should be flushed regularly to remove the sediment at the bottom, which reduces heat transfer. Lamps and light fixtures should be kept clean and dust-free to give maximum light.

Auxiliary devices. A number of devices for modifying heating equipment have been placed on the market with varying claims for their energy-saving capability. Although there is no magic attachment giving ultimate savings, several of these do warrant investigation. An automatic vent damper on a gas or oil-fired boiler stack reduces the amount of heat lost up the chimney after the burner shuts off. A high-pressure burner will increase the efficiency of the flame in an oil-burning furnace. An electric ignition system on a gas furnace will eliminate the need for a constant pilot flame. Automatic thermostats will regulate the temperature to a prearranged pattern and thereby eliminate the possibility of human failure.

Reducing Heat Loss

Heat transferred to the outside through conduction is a major source of heat loss from buildings. Because heat tends to rise, the logical conclusion is that a considerable portion of the loss will be through the ceiling and roof. Heat is lost also to a lesser extent by convection, especially from glass areas and, normally to a very small degree, by radiation. In discussing heat loss from buildings, the term "infiltration" is used frequently. This refers to loss of contained heat through the myriad cracks, crevices and holes inherent in traditional methods of building.

Thermal insulation. Building insulation is designed primarily to combat the transfer of heat through conduction. This is achieved by placing barriers of material with a low heat conductance between the conditioned space and the outside. Entrapped air is a poor conductor of heat and is a cheap commodity; therefore, most building insulations are composed of minuscule pockets of entrapped air.

The measure of insulation value in general use today is the resistance or R-value. Resistance (R) is defined as the reciprocal of conductance. Conductance (C) is the heat (Btu) flowing through a given thickness of material one-foot square at a temperature differential of 1°F. For example, if the conductance value for a given material is determined by experiment to be 0.2, its resistance is 1/0.2 or 5. Such a material would make excellent building insulation, because a four-inch thickness would yield an R-value of 20, equal to the recommended ceiling insulation level for a large section of the United States.

Manufacturers generally designate the R-value obtainable from specific thicknesses of properly installed insulation materials. The optimum amount of insulation expressed as an R-value for a specific locality and type of heating system is available from local energy companies or insulation dealers. Because the proportional reduction in heat transmission decreases relative to the thickness of additional insulating material, a point of diminishing return is reached. Adding insulation beyond this point is futile because, practically speaking, one can never achieve the state of zero heat loss.

The most common insulating materials available commercially today include batts or blankets, pouring, blown, foamed and rigid. Each of these types has useful applications in building construction, although none is a panacea. Batts and blankets are commonly used where the construction is accessible. These generally contain fiber glass or mineral wool. The insulating material is enclosed in paper and comes frequently with a vapor barrier on one side. This vapor barrier may have a shiny metallic surface to reduce transmission by radiation.

Pouring insulation, packaged in bags, also is made of fiber glass or mineral wool, although it may be composed of cellulose, perlite or vermiculite. It is used frequently to insulate existing construction that is inaccessible for installing batts or adding thickness to existing attic insulation.

Blown insulation, composed of fiber glass, mineral wool or cellulose, is installed in walls and ceilings by mechanical force. Access to the walls is gained frequently by cutting a series of holes through the outside face material. Some of the cellulose insulations are damp when installed, and the contained moisture must be expelled before the insulation is fully effective. Professional preservationists have expressed concern about the risk of structural damage resulting from the dissipation of this moisture. The cellulose must be fireproofed to meet government or industry standards, and the fire-retardant chemical used must not be a sulfate compound that will form sulfuric acid in contact with water.

Foamed insulation, composed of plastic material such as urea formaldehyde, is also installed by machinery in the existing wall. Qualified mechanics using specialized machinery are required. Although an efficient insulator, this material was introduced to the construction industry so recently that little is known about the hazards and long-term effects resulting from its use. Like cellulose, the foam gives off moisture after installation that may provoke problems with wall finishes or induce structural decay. There are documented cases of allergic reaction to fumes from the foam. After considering scientific studies linking formaldehyde to cancer, the U.S. Consumer Product Safety Commission voted in 1981 to propose a ban on home insulation containing the chemical. It is difficult and costly to remove the foam if problems are detected after installation.

The rigid types of insulation, composed of cellulose, glass fiber or foamed plastic such as polystyrene or urethane, are available in sheets of various thicknesses and are employed frequently by homeowners for small do-it-yourself jobs because of ease in handling and installation. Care must be exercised to assure a tight fit for maximum efficiency. The plastic boards could present a fire safety hazard unless they are covered with the equivalent of one-half-inch gypsum board.

The goal in any thorough thermal insulation installation is to envelop the conditioned space in an effective wrapping of insulating material. Because the typical conditioned space is defined by a floor, walls and a ceiling, each of these elements must be dealt with as appropriate.

Heat loss through the roof may be remedied readily in most older buildings because the attic is accessible and the joists are open. Insulation is added between the joists to the thickness locally recommended. The insulation may be any of the types previously discussed. If insulation with a vapor barrier is used, it should be installed with the vapor barrier down (adjacent to the conditioned space). When "topping off," or adding to existing attic insulation to increase the R-value, never use a material with an integral vapor barrier or install a vapor barrier between layers of insulation material. The space above the insulation must be adequately ventilated to the outside to allow the escape of moisture.

The thermal insulation of walls presents a number of problems that technology has not solved completely. Where new construction is involved, or the finish material is removed from existing walls, thermal insulation should be installed with a vapor barrier toward the conditioned side. Closed walls are subject to a variety of ills when insulation is attempted. One can never be quite sure how complete the job is, because the installation is made through holes cut into the surface. Settling and shrinkage will produce voids and spots with no insulation whatever. There also is the danger of ruptured walls from improper handling of the gear or excessive pressure build-up. These problems are minor compared to the damage done to the structure by moisture from condensation or by the sulfuric acid formed when some of the cellulose materials are soaked with water. The insulation of sealed walls should be attempted only after all other methods of thermal retention have been used and then only under carefully planned and controlled conditions.

Attempts to insulate walls by adding material to the inside or outside surfaces are not particularly successful. In most cases the losses sustained are not matched by the benefits received. Material added to the inside reduces the size of the room and presents problems to the wall finish, trim, etc. Material added to the outside leads to similar trim problems and seldom adds enough insulation value to be worth the effort. For example, much of the insulation value of aluminum siding is lost when it is installed to allow the ventilation necessary to prevent rot in the underlying wooden structure.

Floors over crawl spaces are most easily insulated with blankets or batts. The insulation material should be held tightly against the floor boards with wire mesh. The vapor barrier must be toward the conditioned side. The surface of the ground in the crawl space should be covered with a 6-mil polyethylene vapor barrier that can be held in place with stones. If the crawl space ventilators are operable they may be closed after freezing weather unless a serious moisture problem dictates that they remain open year-round.

Thermal glass. Thermal glass is available in two forms: fused and banded. Fused glass is composed of two sheets of glass whose edges are heat-sealed. It is available in smaller sizes suitable for domestic glazing and is not as thick as banded glass. Banded glass consists of two sheets of plate glass held in place by a metal edge that forms an airtight seal. This form of thermal glazing is heavier

Library and conservatory, Mark Twain Memorial (1874, Edward T. Potter and Alfred H. Thorp), Hartford, Conn. The Victorian-era use of heavy drapes, low illumination, sun spaces and heavy bookcases helped reduce energy needs. (Photo: Mark Twain Memorial)

and is available in much larger sheets. Both are relatively expensive, although the use of thermal glass in large sheets such as patio doors and picture windows is usually cost effective and certainly adds to the energy efficiency.

Storm sash. In most cases, a more economical method of achieving heat retention at the openings is by installation of storm windows and doors. These may be preengineered metal units or may be custom-milled wood units. The units selected should be compatible in design and construction with the building. Properly designed and installed storm windows will reduce heat loss by conduction, convection and infiltration. Triple glazing is seldom cost effective except under the most rigorous climatic conditions. Many historic houses have been equipped with dismountable vestibules for winter use that are removed during the temperate months. Such a device will drastically reduce infiltration loss at doors with a high use. Houses with entrance halls may present the owner with an opportunity to construct an inside vestibule where none previously existed.

Weatherstripping. Another method of reducing heat loss is through effective weatherstripping. Composed of felt, plastic, rubber or metal, weatherstripping is designed to seal the cracks where an operating unit meets its frame. Old houses often will have no weatherstripping at all, so a complete installation is required. Existing weatherstripping, worn from use, may need overhauling or replacement to be completely effective. Weatherstripping of the proper shape and material

Bottom: Installation of insulation in an old house. (Photo: Jack E. Boucher, Historic House Association of America)

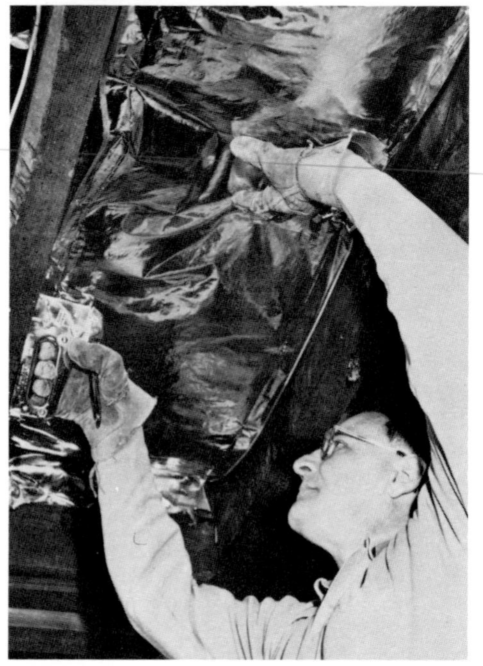

Houses in Fort Conde Village, Church Street East Historic District, Mobile, Ala. Restoration of the Scarpace-Palughi House (1916), far right, included installation of insulation between the interior lath and plaster wall and the exterior wall. (Photo: Carleton Knight III, National Trust)

should be selected for each application.

Caulking. Old buildings tend to "open up at the seams" with the passage of time. Each of these cracks and crevices increases the heat loss by infiltration, and cumulatively they account for a substantial energy waste. All construction joints should be caulked with top-quality acrylic or butyl compound. Cracks too wide to be caulked should be stuffed with oakum or felt. Cracks on the interior around trim, electrical outlet boxes and baseboard also should be caulked.

Pipe insulation. Pipes and ducts carrying conditioned water and air that run through unconditioned spaces are subject to thermal transfer. These should be insulated where they cross attic or crawl spaces. Pipes and ducts in frequently used basement spaces may be acting to temper the surrounding air to a useful degree, and insulating them would be counter-productive.

Alternative Fuels

Much has been published recently about the coming use of exotic fuels such as methane and alcohol. Even so, it probably will be a long time before they play a significant role in filling the energy needs of old houses. More common will be a return to the old standbys, wood and coal. Most older houses were designed for the consumption of these traditional fuels.

Open fires, although psychologically satisfying, are inefficient and, under certain circumstances, dissipate more energy than they save. Nonetheless, a cheery fire in the fireplace can keep a group quite comfortable while a lower temperature is maintained in the rest of the house. During the second half of the 19th century, coal was burned in open grates, and many fireplaces were modified to increase the efficiency of the coal fire. It may be prudent to leave such fireplaces rigged for the effective consumption of coal rather than restoring them to wood-burning capability.

More efficient from an energy conservation standpoint are stoves and furnaces. Airtight stoves properly adjusted can maintain a fire throughout the night on one loading of fuel. Properly located in the home, stoves produce sufficient heat to reduce measurably the consumption of other fuels. The equipment must be installed properly and in good repair. The lethal combustion fumes emanating from a faulty stove have been known to wreak havoc on the unwary.

Energy Audits

Energy audits have become increasingly popular and are available through local energy conservation offices, utilities or private concerns. The purpose of the audit is to determine the efficiency of the total building as a consumer of energy. Typically, the report will include suggestions for modifications to increase that efficiency. Many architects and engineers have attended classes and seminars to qualify them to undertake such work. The procedure may range from a simple on-site evaluation to a complex study involving advanced techniques such as infrared photography. In commissioning an energy audit, one should benefit from the experience of other homeowners in the area and deal only with reputable professionals whose qualifications are established. The results of a properly conducted audit can be invaluable in developing an energy conservation program.

Most of the conservation methods outlined here are simple matters that require nothing more than common sense to put into practice. For the more complex, the homeowner should rely on professional judgment to select the options best suited to the individual structure. The potential exists for sufficient energy to meet our present and future needs. Only sensible stewardship on the part of each of us will enable later generations to make the same statement.

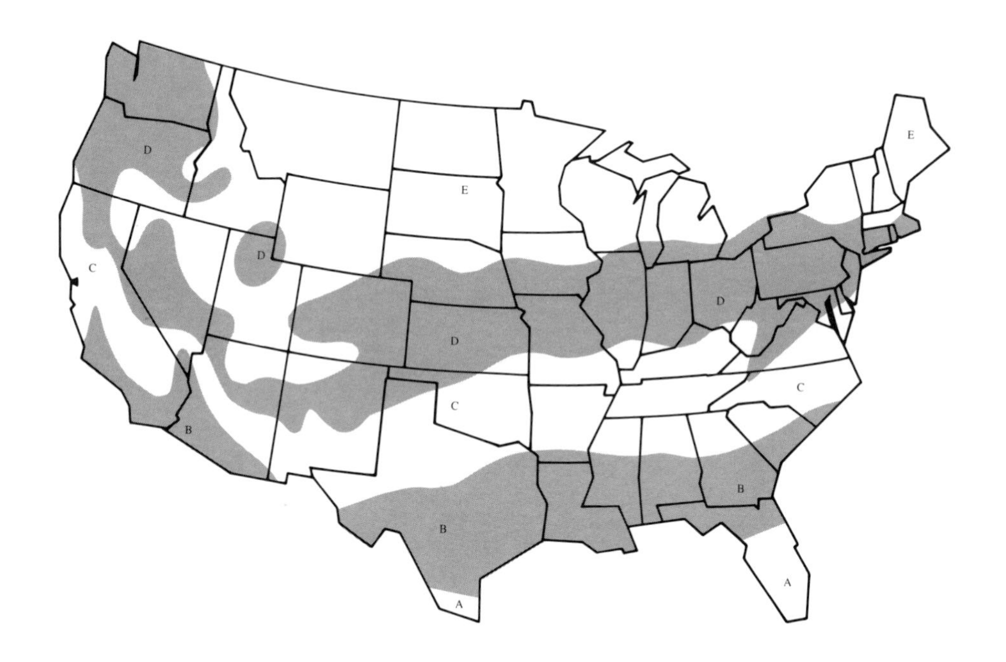

Climate zones of the United States.

How to Save Energy in an Old House

Douglas C. Peterson

To use this table, first locate your proper climate zone (A to E) on the accompanying map. Then use the proper table based on the type of heat. Finally, refer to the measure being considered. The minimum numbers refer to the minimum amount of insulation suggested by any standard. Below these levels you should insulate to the levels recommended by the HUD minimum property standards or slightly better. If your present insulation level falls between the minimum and recommended levels, other energy-saving measures should be considered before additional insulation is applied.

Insulation Requirements

For oil heat, gas heat or heat pump		Climate Zone (see accompanying map)* A	B	C	D	E
Ceilings	*minimum*	R-3†	R-3†	R-6†	R-9	R-9
	recommended	R-19	R-19	R-19	R-30	R-30
Frame walls	*minimum*	none	none	none	none	none
	recommended	none	fill cavity		fill cavity	
Walls of heated basements and crawl spaces	*minimum*	none	none	none	none	none
	recommended	none	none	R-3	R-11	R-11
Floors over unheated spaces	*minimum*	none	none	R-6	R-6	R-6
	recommended	none	none	R-11	R-11	R-19

For electric resistance heat		Climate Zone (see accompanying map)* A	B	C	D	E
Ceilings	*minimum*	R-6†	R-6†	R-9†	R-9	R-11
	recommended	R-19	R-19	R-30	R-30	R-38
Frame walls	*minimum*	none	none	none	R-3	R-3
	recommended	none	fill cavity		fill cavity	
Walls of heated basements and crawl spaces	*minimum*	none	none	none	none	none
	recommended	none	none	R-3	R-11	R-11
Floors over unheated spaces	*minimum*	R-6	R-6	R-6	R-6	R-6
	recommended	R-11	R-11	R-19	R-11	R-19

*Recommended levels are based on the proposed revisions to the HUD minimum property standards, April 1978. The minimum levels in the tables are approximate.
†R-9 if the house has central air conditioning.

How to Save Energy in an Old House

Item	When to Apply this Measure	Percent Reduction in Total Heating and Cooling Bill*	Expected Payback**	Possible Detriment to the Structure
Use by the Homeowner				
Use an entry door opening to an "air lock"	at all times	low	no cost—only savings	none
Close fireplace dampers	at all times between uses	medium	no cost—only savings	none
Lower heat thermostat	anytime it can be done for 2 hours or more	medium	no cost—only savings	none
Adjust cooling thermostat to no more than 10°F below outside temperature	at all times except during periods of extreme humidity	high	no cost—only savings	none
Run only full loads of dishes and clothes	at all times	low	no cost—only savings	none
Keep air-conditioning and refrigerator units free of obstructions	at all times	low	no cost—only savings	none
Turn off appliances when not in use	at all times	low	no cost—only savings	none
Use light bulbs that meet the need—do not oversize	at all times	low	no additional costs	none
Close off rooms not needing heating or cooling	when room can be closed for 5 days or more	low to medium	no cost—only savings	none

*Percent reduction in heating and cooling bill is the probable impact on the present bill: low—less than 5%, medium—5% to 10%, high—over 10% depending on the climate zone, the temperature maintained in the living space and whether it is only heating or heating and cooling combined. Care must be taken when calculating the cumulative effect of adding these reduction percentages. Take the greatest percentage savings first and calculate a new fuel bill; then take the second largest savings and subtract this percentage from the new reduced fuel bill, not the old bill. Proceed in like manner with all improvements until the effect of making the last improvement is understood on the several-times adjusted fuel bill.

**Expected payback is the amount of time necessary to pay back the cost of the investment with energy savings: low—6 to 9 years, medium—3 to 5 years, high—1 to 3 years depending on the installed cost of the item, the climate zone and temperatures maintained in the house.

Item	When to Apply this Measure	Percent Reduction in Total Heating and Cooling Bill*	Expected Payback**	Possible Detriment to the Structure
Mechanical Improvements without Structural Modification				
Furnace-related improvements:				
Flame-retention burner (oil burners)	when burner efficiency cannot be raised above 70% by adjustments to furnace or burner replacement	medium (direct inverse relationship between increase in efficiency and reduction of fuel bill)	high	none
Automatic flue damper (either oil or gas)	all large water or steam units, especially coal conversion units	low	medium	none, but needs correct installation
Barometric draft regulator (oil)	when one is not present or existing one sticks	low	high	none
Water tempering devices (oil or gas)	on forced hot-water systems	medium	high	none
Stack heat recovery units (oil or gas)	when stack temperature of system is at least 550°F after nozzle adjustment and there is a heated space where recovered heat can be ducted	low	medium	none
Ducts and pipes:				
Insulate hot-water heating pipes or cooling pipes	all pipes in unconditioned space—use closed cell foam tubing	medium	high	none
Insulate steam pipes	anytime/everywhere except in heated rooms—use fiber glass tubing only	medium	high	none

Item	When to Apply this Measure	Percent Reduction in Total Heating and Cooling Bill*	Expected Payback**	Possible Detriment to the Structure
Insulate warm air or cooling ducts	anytime duct runs through an unconditioned space—use vinyl-covered fiber glass	medium	high	none
Windows:				
Weatherstrip all contact edges	all windows not painted shut or airtight	medium to high depending on condition	high	none with careful carpentry
Install clamshell locks	install on all windows	low	high	none with careful carpentry
Hot-water heater:				
Insulate jacket of hot-water heater	insulate if tank feels warm to the touch	low	medium	none
Lower temperature setting	set as low as will satisfy needs	low	no cost—only savings	none
Other:				
Caulking around doors and window frames	where parts of house converge—window to wall, door to wall	medium	high	none, helps stop wood deterioration
Water-saving devices for showers, sinks, toilets	anytime, anywhere	low	very high	none
Tempered glass fireplace screen	on fireplace without damper or ill-fitting damper	medium	high	none

Item	When to Apply this Measure	Percent Reduction in Total Heating and Cooling Bill*	Expected Payback**	Possible Detriment to the Structure
Improvements That Reduce Heating Bill, but Could Threaten the House Structurally or Aesthetically				
Wall insulation	see tables on insulation	high depending on climate zone	depends on climate zone and temperature settings	would be detrimental if improperly installed
Attic insulation	see tables on insulation	high depending on climate zone	depends on climate zone and temperature settings	needs proper ventilation
Crawl space insulation	insulate with fiber glass only when crawl space is dry	low to medium depending on temperature and area	depends on climate zone and present temperature settings	in damp areas will trap moisture to wood —do not let insulation reach to floor if termites are problem
Floors over cellar insulation	only when the cellar is 40°F or cooler; see tables	low to medium depending on temperature and area	depends on climate zone and temperature settings	none, but do not leave fiber glass exposed to a living space
Foam insulation	not in old structures	high	medium	moisture problems and possible health hazards
Storm windows and doors	when proper weatherstripping cannot be done	medium	low	aesthetic problems only

Energy Guidelines for an Inner-City Neighborhood

Travis L. Price III

Cities shelter 70 percent of our population. They are the core of our culture and the arena for change in our civilization. As energy prices rise, the drive toward higher densities will probably increase that percentage as a way of conserving both transportation fuels and dwelling energy consumption.

Unfortunately, the decentralized conservation and solar energy movement has somehow become wedded to a decentralized built environment. Rather than adapting itself to the energy-conserving nature of the city, it has spawned itself in the suburban sprawl of single-family dwellings. Needless to say, this will be an inefficient solution for the future solar energy picture. The development of suburban solar sprawl, furthermore, neglects key social responsibilities.

Over the past two decades, urban areas have become a no-growth dumping ground for the poor. The economic and energy needs of the urban poor are increasing to critical levels. Not to address their needs with conservation measures and solar applications first is inequitable and misguided. There have been a number of urban low-income demonstrations over the past five years but their replication is still needed. The three toughest barriers to this replication are the lack of a national conservation and solar policy directed to the cities, the lack of initial funds and the lack of technical knowledge required to adapt energy-saving measures in urban environments.

There seems to be a lack of incentive and knowledge by solar advocates and technicians to address urban energy applications. More important, the alternative energy movement is unaware of the fundamental needs of cities. Understanding these needs and fulfilling them with solar applications will bring solar to the 70 percent of our population that is still skeptical. Clearly, the conservation of buildings, major insulating measures and passive solar applications are natural first steps.

Opposite: Sheffield Street houses in the Manchester neighborhood, Pittsburgh, that will undergo energy-conscious retrofitting. (Photo: Manchester Citizens Corporation)

Cities are plagued by three major needs: jobs, better housing and inflating energy bills. Specific mechanisms exist for tackling these problems. While city officials, planners, bankers and designers are generally providing supportive dialogue, the effective leadership in urban renewable energy applications has tended to come from the advocacy of individuals or community groups.

It is no surprise that the traditional building block of the cities, the neighborhoods, are setting the pace in the use of conservation and solar applications. The neighborhood scale is small enough to set precedents and wield the political power necessary to make them stick. Urban concentration itself, based on common-wall construction and mass transit, is the key energy saver. If a city neighborhood can become energy self-sufficient, the city can follow suit, neighborhood by neighborhood.

The Manchester Demonstration

The Manchester neighborhood of Pittsburgh is setting the pace. Organized and highly productive, the nonprofit Manchester Citizens Corporation has succeeded in tackling a number of neighborhood redevelopment problems. It has secured community control and access to $23 million of Pittsburgh's revenue sharing bonds for housing rehabilitation and new construction. The renovation of more than 1,100 historic structures and the construction of new multifamily housing units are under way in the 79-block area, most of which is listed in the National Register of Historic Places. The Sheffield Street block will be the first block to be rebuilt with extensive energy-saving applications.

With research funding from the Buildings Division of the U.S. Department of Energy, Carnegie-Mellon University in Pittsburgh is establishing a set of inner-city housing energy guidelines specifically tailored for the Manchester neighborhood. Since October 1980, MCC has been using these energy guidelines on the Sheffield block as a demonstration of conservation techniques and solar applications in several historic buildings and in new row house designs. The overall study will be available in workbook format from Carnegie-Mellon University for consumers, designers and financiers in fall 1981, concurrent with results of the Manchester Citizens Corporation demonstration project on the Sheffield block.

To date the most fundamental energy guideline of the study has been to address the existing political-economic context of the city head-on and to adjust the direction of the energy guidelines to serve the neighborhood's needs ahead of all others. Because technical concerns such as energy are not the prime catalysts of change in inner-city neighborhoods, and because community control over a community's development is the prime motivation for citizen involvement, support for the community's role as the key decision maker remains the guiding principle. Energy conservation must start with people. Future replication of any or all parts of the guidelines and demonstration without community participation would be nearly impossible. Without community control and participation, no long-lasting implementation can proceed.

Low-income inner-city neighborhoods in need of development generally tend to be poor, disorganized and inexperi-

Manchester Citizens Corporation staff and a local resident inspect the Sheffield block, where old houses are being rehabilitated and new infill housing built to conserve energy. (Photo: Manchester Citizens Corporation)

Proposals for new energy-saving construction in the Sheffield block, reminiscent of the existing house and porch style found throughout the neighborhood. (Drawings: Urban Design Associates)

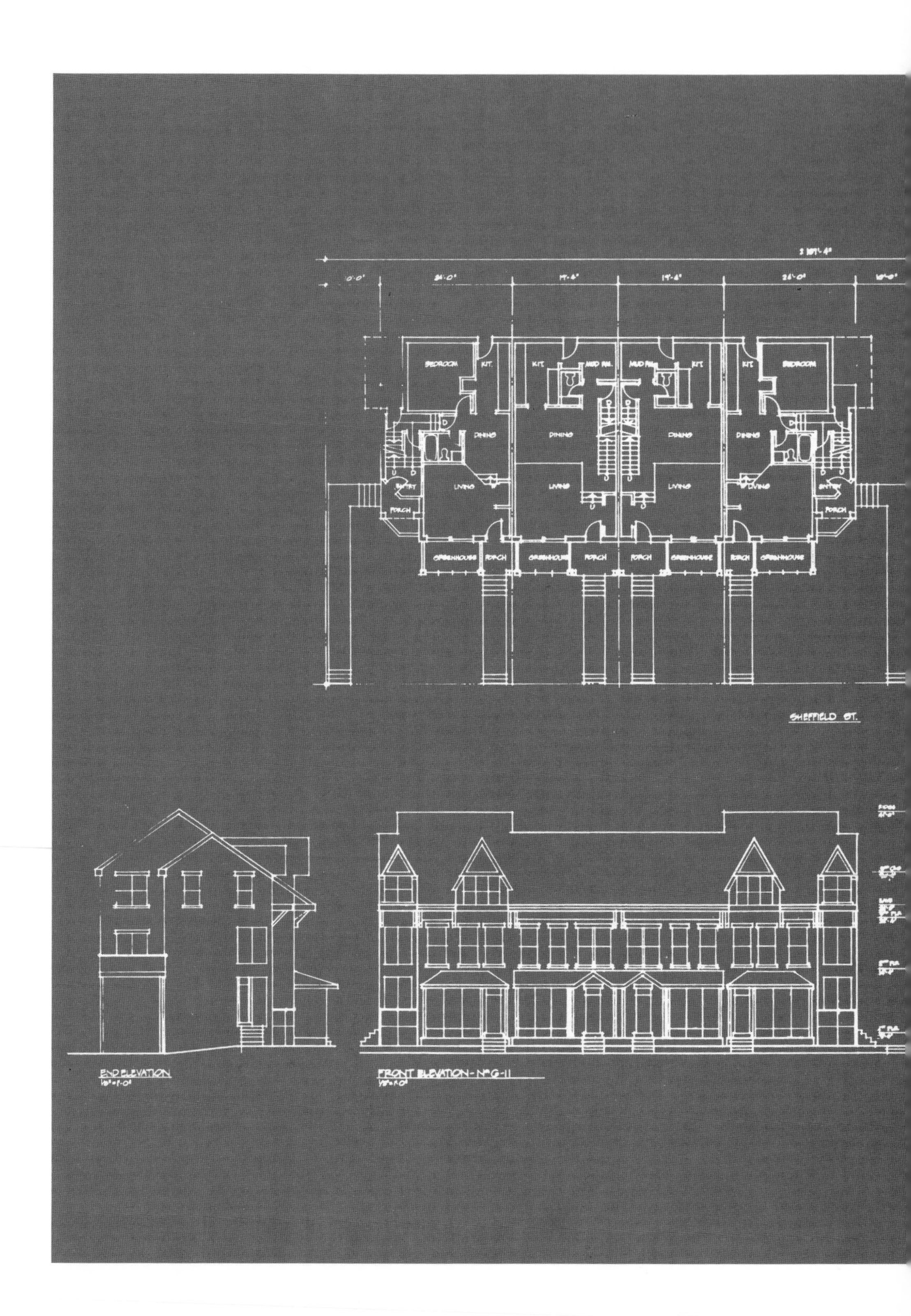

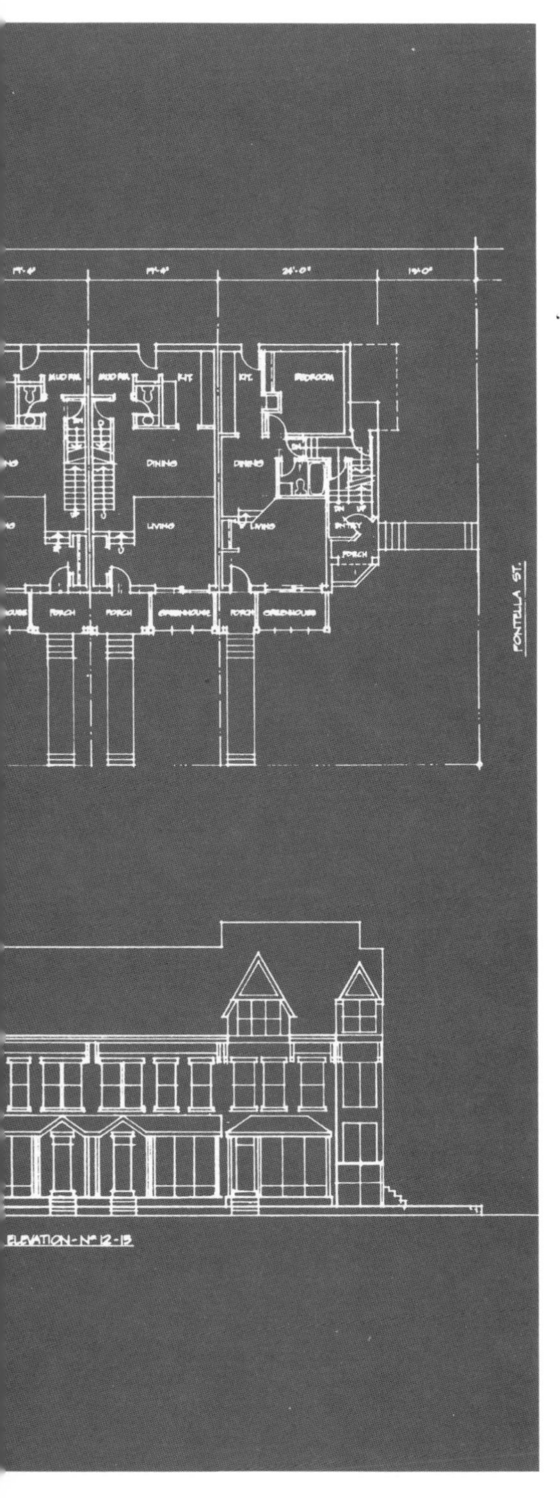

enced. Because of these conditions, technicians (architects, planners, etc.), developers and city government officials from outside the neighborhood tend to distrust the community's capabilities to make well-informed decisions, thereby continuing the present and ineffective decision-making structure from above. In the words of neighborhood activist Msgr. Geno Baroni, under these circumstances, "People who work *for* the community tend to do it *to* the community rather than working *with* the community." The lesson learned is that the community must be in charge and the pace of the decision making must be determined by community needs. Energy technicians tend to find this pace too slow, too erratic or too costly. However, the initial educational slowness is deceiving in that once a common understanding is achieved, the pace tends to pick up and the project's development and eventual replication become assured. Restoration and energy conservation can have a firm foundation if people are considered first.

Concurrent with the neighborhood energy study and the demonstration project, widespread hands-on energy education has brought technical knowledge into the vocabulary of the neighborhood. Without this groundswell of understanding, the entire effort would fail. The energy needs must be understood by each resident before replicative action can take place.

The Energy Guidelines

The development of the technical energy guidelines is structured to analyze the three predominant conditions of the urban housing environment:

1. New infill construction
2. Gut rehabilitation
3. Retrofitting and restoration of occupied dwellings

Much urban housing fits into one of these

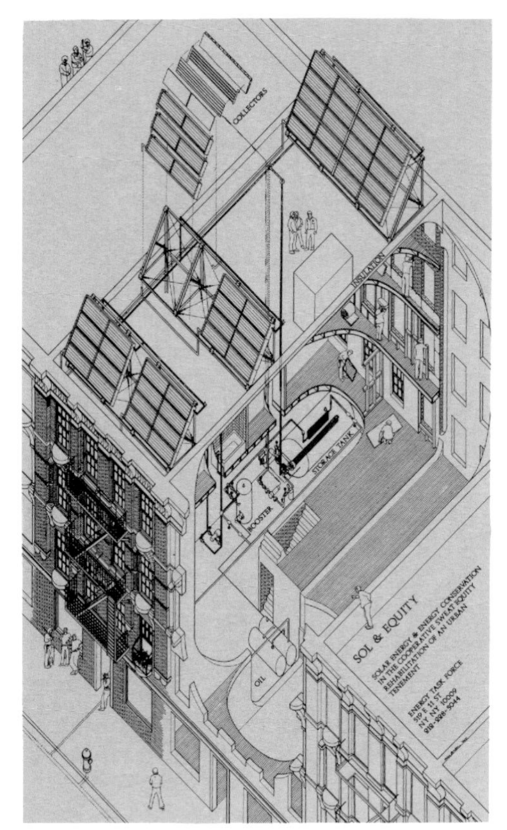

Plan for an innovative rehabilitation of a tenement on the Lower East Side, New York City, in the 1970s with a solar system designed by Travis Price. (Drawing: Henry Dearborn, Energy Task Force)

three categories. The guidelines are the key to the entire study and the energy portions of the MCC demonstration project. The guidelines are the visual, quantitative and written list of design directives for energy-conscious inner-city development. To develop these guidelines, the overall studies are organized into two distinct categories: load and supply. Reducing the energy loads and increasing renewable supplies are the major goals.

Load. The load portion consists of all the data-gathering and analysis necessary to establish the base energy requirements of the Manchester neighborhood and the Sheffield block. With the energy load requirements profiled, the baseline of energy saved by implementing the design guidelines is established. The load data account for all the energy streams entering the neighborhood and determining where and how energy is used or lost. Understanding energy needs is crucial before finding new energy supplies.

Supply. The supply portion analyzes all aspects of energy-conscious design alternatives quantitatively by Btu per square foot per dollar. The supply alternatives include all aspects of energy efficiency, energy conservation and on-site renewable energy production. Optimal design options are established by the full economic integration of these options. The detailed design for the demonstration for Sheffield Street will calibrate the overall potential impact of the energy guidelines for the entire Manchester neighborhood.

The guidelines are integrated into the Sheffield demonstration project, and their predictive savings compared to the actual savings of the finished design are to be analyzed. Key reports developed for the study data base are: a building inventory of the entire neighborhood, energy audits, user profiles, waste management studies, urban agriculture alternatives, utility integration, on-site cogeneration potential, historic preservation guidance, population projections, existing standards of urban housing construction and a comparison with new suburban construction and energy costs.

The load and supply analyses quantitatively integrate the full energy picture, providing the best design options in a number of high-density building configurations. Two key design approaches currently recommended are:

1. Conservation (increasing insulation)
2. Direct gain (using direct southern sunlight)

These two options are analyzed for all three high-density conditions: new, rehab and retrofit. Architecturally, the best direction is the conservation solution with direct solar gain for retrofit, rehab and new construction. Proper use of nighttime insulation over the glazing (i.e., R-5 shutters) is imperative. Otherwise, a superinsulated south wall is necessary.

Critical to the overall energy guidelines is the full integration of these architectural options into the neighborhood energy needs and the historical restrictions for the neighborhood. The relationship with the utility grid and the development of a block or neighborhood power generation station are also examined with the energy design options.

An overall energy plan based on three future scenarios reflects the detailed quantification of the Sheffield block demonstration. The three scenarios predict the energy consumption of the block. The first assumes no energy-conscious efforts. The second gives a moderate-conservation scenario of using cost-effective measures. The third assumes no dependence on nonrenewable resources other than embodied energy in the building materials.

Importantly, the energy plan and demonstration are useless without a financial strategy and the backing to implement it. In the urban context, especially impoverished neighborhoods, the key issue is a lack of capital along with the more pressing problem of leaking roofs and deteriorating porches.

The bottom-line economic guideline of the study assumes that each urban dweller, either renter or owner, cannot afford any additional debt service, be it energy costs or mortgage costs. The first payments for an energy-saving measure must be no greater than the money saved by taking that measure.

The guidelines developed for Pittsburgh adhere to this economic formula. This formula solidifies the replicability of the demonstration and the guidelines in other neighborhoods. It is critical in the guidelines that simplicity and economic validity be integrated. Further conformance of the energy economic analysis to the community's financial capabilities has maintained the economic solidity of the guidelines development.

Savings on energy bills for historic buildings rehabilitated with energy-conscious design can equal 65 percent over conventionally restored buildings. Likewise, new construction adhering to historical concerns can achieve an 85 percent reduction in operating energy costs. Both of these figures assume proper consumer and designer education. Both also are accomplished without increasing the buyer's debt service the first year.

The urban neighborhood is a microcosm of the entire energy picture. Existing neighborhoods are ready for energy-conscious redesign. The precise and best solutions will vary from neighborhood to neighborhood and from climate to climate. It is clear that options using conservation and passive solar applications have a substantial effect on urban neighborhood energy consumption.

Energy-conscious guidelines for cities may well become the crucial energy answer for the remainder of the century. Certainly, without them historic buildings in cities may find themselves restored but unoccupied.

2. Old Buildings as Energy Investments

ANNO DOMINI MDCCCXCVI
GSA
Renovation
Old Post Office

Assessing Energy Conservation Benefits: A Study

Calvin W. Carter

When the rehabilitation of old buildings is subjected to scientific analysis and computation, there is no doubt that it saves more energy than does new construction.

This is the conclusion of a study, *Assessing the Energy Conservation Benefits of Historic Preservation: Methods and Examples*, issued by the Advisory Council on Historic Preservation in spring 1979. This study should profoundly influence the preservation movement and perhaps revolutionize the way effects on the built environment are evaluated. Its influence already is being felt in several areas, but its basic concept has yet to achieve full recognition and support.

Quantifying Energy Significance

The need for the study grew out of council responsibilities under section 106 of the National Historic Preservation Act of 1966. This section requires federal agencies that undertake, assist or license activities that adversely affect properties listed in or eligible for the National Register of Historic Places to seek council comments. Frequently, cases of adverse effect involve proposals to tear down historic buildings and replace them with new construction. When energy savings became a national priority, the council wanted to be able to make enlightened judgments not only on historical significance and social and economic factors, but also on the energy trade-offs in these cases.

The council also looked on the study as a way to help meet its responsibilities under the Public Buildings Cooperative Use Act of 1976. This legislation directs the U.S. General Services Administration to lease or to purchase and rejuvenate buildings of historical and architectural value for federal office space and other mixed uses and requires the agency to seek council assistance in identifying "suitable" properties. The energy formulas called for by the study were another tool to assist in determining this.

The particular section 106 case that served as the impetus for the study concerned Lockefield Garden Apartments (1935) in Indianapolis. One of the first public housing projects in the country and a prototype for many others, this complex was threatened by demolition with new housing planned nearby. Demolition required permission from the U.S. Department of Housing and Urban Development, which asked the council to comment at its May 1977 meeting.

At that time, the council was aware of the work done by Richard G. Stein, FAIA, and the Center for Advanced Computation, University of Illinois, and saw this as an opportunity to bring forth the concept of embodied energy as a part of the argument for retention of the property. The calculations were crude, but the point was made. The council was able to show that the energy saved by reusing the complex would be at least enough to heat 500 single-family homes for five years.

Subsequently, it was decided that the approach should be refined and easily used formulas developed. The research proposal asked for formulas to measure:

1. Energy already existing in a structure to be rehabilitated
2. Energy needed for construction and rehabilitation
3. Energy needed for demolition and preparation of a construction site

Opposite: Old Post Office (1892-99, Willoughby J. Edbrooke), Washington, D.C., which embodies 2.9 million gallons of gasoline. (Photo: Lillian M. O'Connell, U.S. General Services Administration)

Razing of the old Panamanian legation (1893), Washington, D.C. Demolition wastes embodied energy as well as destroying buildings. (Photo: John J.G. Blumenson, National Trust)

4. Energy needed to operate a rehabilitated or newly constructed building

In addition, three case studies to test the new formulas were requested: one of a housing development (Lockefield Gardens was stipulated), one of a commercial complex and one of a family residence in the Washington, D.C., area. Booz, Allen and Hamilton, Inc., of Washington, D.C., was selected to undertake the study, which began in fall 1977.

Energy Concerns

Background material noted that despite the effects of the 1973 embargo, oil consumption in the United States had grown at a rate of 3 percent each year, an increase from something more than 6 million barrels a day in 1973 to 8 million a day in 1977. This increase is supported by imported oil. In 1973, 36 percent of the total supply came from foreign sources; by 1977, the amount had reached 47 percent, and it is still rising. Consequently, the United States now is more vulnerable to an energy crisis than it was six years ago. The National Energy Act, passed by Congress and signed into law in fall 1978, specifies a number of ways to cut back use of imported oil.

Buildings account for a substantial amount of the energy consumed in the United States, and the efficient use of energy in this area has become a major national goal. Comparisons show the significance of consumption in the building sector. Industry is the greatest consumer of energy, using 43 percent, but buildings are second—accounting for 32 percent of the total. Transportation follows with 25 percent.

Most of the energy consumed in buildings provides for human comfort. Heating is the major item in both residential and nonresidential buidings, while lighting is an important factor only in nonresidential use. A significant amount of

Demolition Energy for Existing Buildings (Concept Model)

Construction Type	Small Building (5,000-15,000 sq. ft.)	Medium Building (50,000-150,000 sq. ft.)	Large Building (500,000-1,500,000 sq. ft.)
Light (e.g., wood frame)	3,100 Btu/sq. ft.	2,400 Btu/sq. ft.	2,100 Btu/sq. ft.
Medium (e.g., steel frame)	9,300 Btu/sq. ft.	7,200 Btu/sq. ft.	6,300 Btu/sq. ft.
Heavy (e.g., masonry, concrete)	15,500 Btu/sq. ft.	12,000 Btu/sq. ft.	10,500 Btu/sq. ft.

Source: *Assessing the Energy Conservation Benefits of Historic Preservation: Methods and Examples*, by Booz, Allen and Hamilton for the Advisory Council on Historic Preservation, January 1979.

energy also is used each year to construct new buildings and preserve old ones. As the population grows, more residential, commercial and institutional buildings are needed, and economists predict that building activity will increase steadily over the next five years, despite temporary setbacks.

The energy used to construct a building involves more than just the fuel to run the bulldozers and cement mixers. To get a proper idea of how much energy goes into a building, it is necessary to take into account the embodied energy of materials. Embodied energy comprises the amount of energy it takes to produce and deliver the materials. This value—always measured in British thermal units (Btu's)—varies greatly among different materials, depending on the amount of processing required. For example, take a five-ton steel girder delivered to a construction site. The energy invested in it from the time the iron is mixed through the various stages of processing and fabricating the steel is 257 million Btu's. Transporting the finished product to the site and installing it in the building requires 13 million Btu's. Thus, the embodied energy in a five-ton steel girder totals 270 million Btu's, the amount of energy contained in 2,000 gallons of gasoline.

Energy Costs to Demolish and Replace Selected Historic Buildings

	Gallons of Gas
St. Louis, Mo.	
Wainwright Building	6,454,200
Syndicate Trust/Century Building	8,988,900
Laclede Gas Building	1,464,000
Old Post Office	3,662,000
Washington, D.C.	
Andrew Mellon Building/ National Trust Headquarters	776,100
Christian Heurich Mansion	209,600
Renwick Gallery	343,700
Decatur House	112,700
Hay-Adams Hotel	804,900
Willard Hotel	3,125,900
Old Post Office	2,913,000
Pension Building	2,022,800
Union Station	7,432,300

Portion of the Lockefield Garden Apartments (1935), Indianapolis, Ind. (Photo: Historic Landmarks Foundation of Indiana)

It is obvious that the nation's stock of buildings represents a major investment. In fact, it is the major energy investment of our culture. Replacing all the existing buildings in the United States would require the world's entire energy output for one year—approximately 200 quadrillion Btu's of energy. On this basis, it would seem to be advantageous to preserve buildings as long as they can be useful.

Case Studies

To determine whether this assumption is true and whether preservation really saves energy, three major preservation projects were studied.

The Lockefield Garden Apartments, one of the three, had been abandoned for several years and was slated for demolition. It was found that the apartments represented a total energy investment of 570 billion Btu's, which is equivalent to the energy contained in 100,000 barrels of oil or 4.5 million gallons of gasoline.

An alternative to new construction, of course, is to preserve buildings, restoring them for modern use. Such renovation actually can save energy. A good example of this approach is the Grand Central Arcade in the Seattle Pioneer Square Historic District. Built in 1899 as a hotel and restored in 1972 for offices and shops, the Grand Central Arcade contains 80,000 square feet of usable space, reclaimed at a cost of only 17 billion Btu's—5 billion to manufacture the materials needed to accomplish restoration and 12 billion to put them in place. A new building with comparable floor space constructed of modern materials would require an investment of 109 billion Btu's, 85 billion for materials and 24 billion to fashion them into the completed structure. The energy savings in this case is 92 billion Btu's, the equivalent of 730,000 gallons of gasoline, enough to power 250 automobiles for 60,000 miles.

Another example on a smaller scale was the Austin House, an adapted carriage house on Capitol Hill in Washington, D.C. Extensive renovation has turned the Austin House into a three-unit apartment building. Only the outside shell was left intact. Even so, preservation was the winner. While it took 370 million Btu's to provide the materials for renovation, 1,430 million Btu's would have been required to provide the materials for new construction. The energy needed to put the materials in place is the same in both cases—270 million units. Thus, rehabilitation saved more than 1,000 million Btu's.

It is important to note that existing buildings can be rehabilitated so that they consume amounts of energy similar to today's energy-conscious buildings, or even less. Many old buildings have features such as massive walls and efficient ventilation systems that can make them more efficient than recently built structures.

The council study shows that it will take 117 million Btu's to heat and cool the Austin House for a year, but it will require 122 million Btu's to heat and cool an equivalent new building. Thus, the Austin House is 5 percent more energy efficient than a comparable new structure.

Adding the Btu's saved in restoration to those saved in the operation of the structure over its expected life, it is found that the Austin House will conserve 1,220 million Btu's, the amount needed to heat another three-unit apartment building for a decade.

The Grand Central Arcade in Seattle, converted to offices and shops a year before the consciousness-raising oil embargo and before the development of present energy conservation techniques, consumes only about 6 percent more than an equivalent new building, using 6.2 billion Btu's a year. This is especially remarkable in view of the huge open areas found in the structure.

The slight operational energy increase is offset by the great initial energy savings from preservation—92 billion Btu's. If the amount of energy needed to operate a new building each year—5.8 billion Btu's—is multiplied 16 times, it equals 92 billion, the amount of energy saved by the retention and renovation of the Grand Central Arcade. Thus, preservation saved enough energy to heat and cool a comparable new building for 16 years.

Grand Central Arcade (1899), Seattle, preserved for reuse as shops and offices (1972, Ralph Anderson, Koch and Duarte). (Photo: Art Hupy)

Austin House, Capitol Hill, Washington, D.C., converted to apartments. (Photo: William I. Whiddon)

Computation of Savings

These figures were obtained using methods developed as part of the study. The techniques can be used in any city, by any group, to determine the energy savings of a proposed preservation project.

The first and simplest of the methods is the Building Concept Model. A minimum of information gives a rough estimate of the energy to be saved. Consider a turn-of-the-century warehouse in an urban setting. If the gross floor area is known, the total amount of energy embodied in the entire structure can be estimated. Using a table that gives a per-square-foot figure for various categories of buildings, we can determine that the energy rating for warehouses is 560, or 560,000 Btu's per square foot. The rating for office buildings is 1,640, or 1.64 million Btu's per square foot. Thus, a building with 80,000 square feet of floor space, such as the Grand Central Arcade, represents an energy investment of roughly 131 billion Btu's.

To compute a more accurate estimate of energy savings, the Building Survey Model is used. This model requires more information about the structure, but the necessary information is obtained easily. In addition to the floor area, the approximate quantities of seven primary materials used in the building must be known. For example, if the structure were made of brick, the entry for stone and clay prod-

Embodied Energy of Building Types (Concept Model)

Building Type	MBtu/sq. ft.
Residential: 1-family	700
Residential: 2–4-family	630
Residential: garden apartment	650
Residential: high rise	740
Hotel/motel	1,130
Dormitories	1,430
Industrial	970
Office	1,640
Warehouses	560
Garages/service stations	770
Stores/restaurants	940
Religious	1,260
Educational	1,390
Hospital	1,720
Other nonfarm buildings	1,450
Amusement, social and recreational	1,380
Miscellaneous nonresidential	1,100
Laboratories	2,070
Libraries, museums, etc.	1,740

Source: *Energy Use for Building Construction,* by Energy Research Group, Center for Advanced Computation, University of Illinois, and Richard G. Stein and Associates, December 1976. Reprinted in *Assessing the Energy Conservation Benefits of Historic Preservation.*

Energy Embodiment of Primary Building Materials (Survey Model)

Material	Embodied Energy per Unit
Wood products	9,000 Btu/board ft.
Paint products	1,000 Btu/sq. ft. (450 sq. ft./gallon)
Asphalt products	2,000 Btu/sq. ft.
Glass products	
Sheet-glass window	15,000 Btu/sq. ft.
Plate-glass window	40,000 Btu/sq. ft.
Stone and clay products	
Concrete	96,000 Btu/cubic ft.
Brick	400,000 Btu/cubic ft.
Primary iron and steel products	25,000 Btu/lb.
Primary nonferrous products	95,000 Btu/lb.

Source: *Assessing the Energy Conservation Benefits of Historic Preservation.* Approximations based on data for a variety of products included in *Energy Use for Building Construction.*

Embodied Energy of Paint Products (Inventory Model)

Material	Embodied Energy per Unit
Exterior water-type trade sales paint	489,000 Btu/gallon
Interior water-type trade sales paint	437,000 Btu/gallon
Interior oil-type trade sales paint	508,000 Btu/gallon

Source: *Energy Use for Building Construction.* Reprinted in *Assessing the Energy Conservation Benefits of Historic Preservation.*

Embodied Energy of Wood Products (Inventory Model)

Material	Embodied Energy per Unit
Dressed softwood boards	7,900 Btu/board ft.
Dressed hardwood lumber	9,700 Btu/board ft.
Hardwood flooring	10,300 Btu/board ft.
Shingles and handsplit shakes	7,300 Btu/sq. ft.
Double-hung wood window unit	1,127,000 Btu each
Other wood window units	1,830,000 Btu each
Interior and exterior panel door	873,000 Btu each
Hollow-core door	347,000 Btu each
Solid-core door	1,191,000 Btu each
Wood garage doors	3,321,000 Btu each
Finished wood molding	18,000 Btu/board ft.
Sawn wood structural member	6,400 Btu/board ft.
Glued laminated wood structural member	16,700 Btu/board ft.

Source: *Energy Use for Building Construction.* Reprinted in *Assessing the Energy Conservation Benefits of Historic Preservation.*

ucts would be consulted. The value for brick is 400,000 Btu's per cubic foot. Using this factor, the analysis shows that 12,000 cubic feet of brick in a building represents an energy investment of 4.8 billion Btu's. The types of heating and cooling systems and the kinds of fuel used in the building by its mechanical systems complete the calculation.

The third computation method is the Building Inventory Model, a detailed analysis that provides the most accurate idea of the amount of embodied energy in an existing structure. The basic calculations are the same as in the two previous models, but there are more of them. Additional data are required—the climate, building characteristics, patterns of energy use, etc. For instance, it is possible to go so far as to calculate the energy in the paint needed for the interior of the renovated building. The entry for interior oil-type trade sales paint products shows an energy value of 508,000 Btu's. This level of detail may or may not be appropriate for a particular project, but it is important to know that this type of analysis is available, if needed, to show conclusively that preservation can save energy.

Goal: Federal Policy

Still, even federal policy does not yet take the embodied energy concept into account. The council has adopted recommendations on ways to do this and has transmitted them to the president and to Congress. Proposed means of accomplishing this objective are:

1. To amend section 10 of Executive Order 11912, "Energy Policy and Conservation," to include embodied energy and demolition energy in the equation developed for estimating and comparing the life-cycle costs of federal buildings.
2. To further amend section 10 of the same executive order to permit agencies that meet needs for new space through

rehabilitation rather than new construction to apply the energy credits earned (determined with the analysis developed by the council) against the 20 percent reduction in energy consumption they are required to make by 1985.

3. To incorporate the energy analysis developed by the council into the environmental impact statement process for evaluating projects involving new construction or rehabilitation.

A strong historic preservation program means a strong energy program, and with the saving and renewal of the present building stock, it will be possible to meet the challenges of the 1980s. Preserving existing buildings means preserving existing energy. Helping to save buildings means helping to build a strong nation.

Restoration of the Winder Building (1847-48), Washington, D.C., for federal agency use. In the background is the Old State, War and Navy Building (1871-88, Alfred B. Mullett), now used for White House offices. (Photo: U.S. General Services Administration)

The Concept of Embodied Energy

William I. Whiddon

Embodied energy—the amount of energy required to produce materials used in building construction and to put them in place—represents a significant portion of America's annual energy use. Construction of new buildings in the United States accounts for more than 5 percent of the total U.S. energy use each year. On the average, 15 percent of the energy required for building construction, which is equivalent to 100 million barrels of oil, is used directly on the job site, for fuels to run equipment and by the people involved in the labor. The remaining 85 percent is embodied energy, equivalent to 500 million barrels of oil a year in the United States—roughly enough energy to fuel all of the automobiles in the Washington, D.C., area for nearly 20 years.

Historic preservation has the potential for displacing a large fraction of the energy used directly at the job site and embodied in construction materials. As new and existing buildings are made increasingly efficient in the ways they use energy, embodied energy becomes an even more significant fraction of the energy investment required in the use of buildings. For example, the U.S. General Services Administration has set 100,000 Btu's as a standard of energy use for new federal office buildings. That is less, comparatively, than a gallon of gasoline per square foot per year for heating, cooling and lighting. Comparing that annual energy use to the energy required to build a typical square foot of new office space, the amount of energy initially invested in a building—equivalent to about 12 gallons of gasoline per square foot—is enough to heat, cool and light the same building for more than 15 years. That ratio will increase, and embodied energy will become more and more a factor in energy use in the United States.

Opposite: New York State Capitol (1867-99, Thomas Fuller, H.H. Richardson, Leopold Eidlitz, Isaac Perry), Albany, embodying 7.8 million gallons of gasoline. (Photo: Walter Smalling, Jr., U.S. Department of the Interior)

Some energy use can be measured and tracked by devices as simple as fuel bills on a construction site or the direct energy use in a house. Estimates of embodied energy, however, have been difficult to derive and are not directly measurable.

Advisory Council Study

The study conducted by Booz, Allen and Hamilton, Inc., for the Advisory Council on Historic Preservation was based on the pioneering research of Richard G. Stein, FAIA, at the Center for Advanced Computation, University of Illinois. Stein's work was sponsored by the Office of Buildings and Community Systems, U.S. Department of Energy, and resulted in the publication in 1980 of the *Handbook of Energy Use for Building Construction*. The most recent version of the handbook is a tabulation of the embodied energy of hundreds of building products and appliances.

Stein developed a complex energy input-output model for the building industry, which was used to derive estimates of the embodied energy for building materials. The existing data base includes information only for construction activities occurring in 1967. Energy use patterns for the construction industry before or after 1967 have not been precisely determined. Since 1973, there has been a significant effort to change industrial processes, which is likely to change slightly some of the values computed for embodied energy.

The estimates of energy embodied in old buildings developed with this data base do not reflect the energy actually used to produce the original materials in existing buildings, but rather the energy that would be required for comparable materials if the building were to be constructed today.

Computation of Embodied Energy

To find a simple way to estimate the energy value of existing buildings, Booz, Allen and Hamilton developed three increasingly detailed methods for the Advisory Council on Historic Preservation. The first method, the Building Concept Model, directly compares existing buildings with new buildings by building type. It is the crudest estimate of embodied energy, essentially computing the embodied energy in a contemporary building and using that as a basis for estimating the embodied energy in a historic building. The second method, the Building Survey Model, is a quick, although admittedly rough, analysis of energy in primary building materials used in a given building. It is better than the first method simply because material differences can be easily taken into account. The third method, the Building Inventory Model, is a detailed procedure allowing the user to be as precise as available information will allow in figuring the amount of energy embodied in a building.

It is interesting to note that residences of various types, from single-family units to duplexes and garden apartments to high rises, all seem to require a similar amount of energy for construction and materials. Hospitals are one building type that requires a high amount of energy per square foot of building space, and offices follow closely behind. The urban environment is largely composed of residences and offices, and many of the historic buildings to be preserved are going to be used for those purposes. There appears to be a high potential for reducing requirements for energy through preservation while meeting the demands of people for housing and offices.

Austin House, Capitol Hill, Washington, D.C., one of three buildings studied for the Advisory Council on Historic Preservation embodied energy survey. (Photo: Marcia Axtmann Smith)

Detail of Austin House carriage house and courtyard. (Photo: William I. Whiddon)

Examples of Methods and Results

The concept method for comparison of old with new construction requires equating the old building with a new building of the same type and size. For example, the embodied energy of a typical 10,000-square-foot office building is equivalent to 12 gallons of gasoline per square foot, a total of 120,000 gallons of gasoline. The Austin House in the Capitol Hill area of Washington, D.C., was used as a case study in the council work. It is a 2,700-square-foot carriage house converted to a small apartment building. To calculate the embodied energy, simply multiply the 2,700 square feet by the 600,000 to 630,000 Btu's per square foot required for a three-unit apartment building. The resultant 1,700 million Btu's of embodied energy is equal to more than 12,000 gallons of gasoline.

The survey method assesses the embodied energy in the actual materials employed in the existing buildings being studied. Embodied energy in materials varies over a wide range, for example:

Paint, an energy-intensive product, represents 450,000 Btu's per gallon, while a gallon of gasoline will produce only 125,000 Btu's. If paint typically covers 450 square feet per gallon, energy embodied in a painted surface is approximately 1,000 Btu's per square foot for each coat of paint.

Stone and clay products, used in great bulk in both old and new buildings, are also energy intensive, ranging from 100,000 to 400,000 Btu's per cubic foot of material.

Metals range from steel, rated at 25,000 Btu's per pound, to the nonferrous products, such as aluminum, at up to 95,000 Btu's per pound.

The Austin House embodies about 1,600 million Btu's of energy in primary materials, equivalent to about 12,000 gallons of gasoline. The first estimate, using the concept method, was 1,700 million Btu's, which is within about 10 percent of

Porch on Austin House, now used as a small apartment building. (Photo: William I. Whiddon)

Embodied Energy of Rehabilitation vs. New Construction

Project Size	Rehabilitation	New Construction
Lockefield Garden Apartments (517,000 sq. ft.)	120,000 MMBtu	350,000 MMBtu
Grand Central Arcade (80,000 sq. ft.)	17,000 MMBtu	109,000 MMBtu
Austin House (2,700 sq. ft.)	640 MMBtu	1,700 MMBtu

Source: *Assessing the Energy Conservation Benefits of Historic Preservation: Methods and Examples*, by Booz, Allen and Hamilton for the Advisory Council on Historic Preservation, January 1979.

Austin House: Embodied Energy of Rehabilitation Materials (Survey Model)

Material	Quantity	Embodied Energy per Unit	Total
Wood	3,078 board ft.	9,000 Btu/board ft.	28 MMBtu
Brick	304 cubic ft.	400,000 Btu/cubic ft.	121 MMBtu
Concrete (plaster)	720 cubic ft.	96,000 Btu/cubic ft.	69 MMBtu
Windows	388 sq. ft.	15,000 Btu/sq. ft.	6 MMBtu
Insulation	4,472 sq. ft.	8,000 Btu/sq. ft.	36 MMBtu
Surveyed materials subtotal (70% of total rehabilitation materials)			260 MMBtu
100% of embodied energy of rehabilitation materials			370 MMBtu
Rehabilitation construction energy			270 MMBtu
Total for rehabilitation			640 MMBtu
Total for new construction (2,700 sq. ft. x 0.63 MMBtu/sq. ft.)			1,700 MMBtu

Source: *Assessing the Energy Conservation Benefits of Historic Preservation.*

Molly Brown House (1887-92), Denver, a medium-sized historic house that represents an energy investment of 42,583 gallons of gasoline. (Photo: Historic Denver)

this more precise computation using the survey method.

The third, or inventory, method of measuring embodied energy requires a detailed materials analysis, but the simple arithmetic summations can be done easily. There is some question about whether or not the detail and precision of this method are necessary for the kinds of analysis and arguments justifying historic preservation, but certainly it can be used for specific projects to measure precisely the total energy value of a building.

Implications for Preservation

The Austin House case study suggests general conclusions on the potential of preservation as an energy conservation technique. The building was completely gutted and rehabilitated; the shell of a carriage house was converted into something quite different. Although a large amount of material was used, the rehabilitation required only 40 percent of the energy needed to construct a new three-unit apartment building.

This result has significant implications for preservation and energy conservation in this country. The potential energy savings that can be realized from recycling buildings may match the savings that the U.S. Department of Energy and various government agencies expect to see realized by new forms of energy in the future.

If this single example is representative, the rehabilitation of two million single-family residences could result potentially in energy savings equivalent to 172 million barrels of fossil fuel. To give a brief comparison, the Department of Energy would like to see two million passive solar buildings erected by 1986; this is expected to save one-tenth of a quad (10^{15} Btu's) of energy. Preservation of a like number of dwellings presents an opportunity for 10 times the immediate energy savings, although there are not the continuing savings that come with new passive solar buildings.

Results such as this indicate that preservation and energy conservation can indeed be considered synonymous. Preservation today has the potential to become an instrument of national energy policy.

Using the Embodied Energy Argument in Local Planning Controversies

James Vaseff, AIA

At a time when the industrialized world is learning quickly that the energy resources of our planet are finite, the concept of embodied energy is beginning to stand as one of the philosophical piers of historic preservation. But advancing the concept of embodied energy by itself is not terribly effective in calming a local preservation-planning controversy. It is similar to trying to explain how recapturing petrodollars with exported Pacific Northwest lumber will benefit a mom-and-pop deli. Obviously, conservation of fossil fuels is a national goal, but the nature of property development in the United States (primarily by private capital) limits the vision of each project to its property lines and the bottom line of yearly cash returns. Instead of launching a defense of preservation based on embodied energy, the concept is best introduced in support of other, more expedient, issues.

The increasing value of the embodied energy argument can be understood by analyzing it in terms of preservation and economics. These two issues have had, at best, a symbiotic relationship. They also have changed dramatically in the past several years. For this discussion, preservation is defined as local advocates' efforts to encourage recognition and use of the historic environment. Economics is manifested in capital and energy.

Two Perspectives on Preservation

Those who have been involved with preservation for the past decade or two have seen it progress from a "heartstrings" effort to a sophisticated movement with resourceful strategies. In the most optimistic light one could say that preservation is no longer an isolated interest; it is part of the action: The National Register of Historic Places is a planning tool; the Tax Reform Act of 1976 and related legislation offer tax incentives for commercial rehabilitation; local landmark and design review commissions have been initiated throughout the country; preservation law is a recognized discipline with able practitioners. Preservation's image over the past decade in the eyes of municipalities and developers has matured from that of an *enfant terrible* to a sophisticated and responsible (in most cases) advocate.

Unlike preservation, capital has no sentiment. People believe that their capital must grow, and there are many players —private, public and institutional—who are out to make the most of their resources. Although many vested interests operate in the economic development arena, the most visible in a local controversy is usually the property developer. The developer is invariably associated with a structure (new or old) that has become the local issue. A look at the situation is necessary to understand why developers, to the preservationists' perspective, have such limited vision.

When energy and money were plentiful (35 cents per gallon for gasoline and 7½ percent interest rates) the idea of reuse was, in the eyes of the developer, a luxury. It is now known that construction costs for adapting old buildings to new uses can be much lower than new construction, but before 1976 there were tax disadvantages in reusing old structures; public investment in roads, sewers and utilities was almost totally directed toward constructing strip development; zoning and building codes did not recognize reuse; recalcitrant banks did not loan money for old buildings. A mobile

Opposite: Federal Home Loan Bank Board (1977, Max Urbahn Associates), Washington, D.C., built on the site of a historic bank before energy became a preservation argument. (Photo: U.S. General Services Administration)

market followed, by auto, commercial development wherever it was located. To go into a part of town that, for many people, was undesirable and to wrestle with the cited hurdles in developing a property took a developer who would personally cover the extra expenses that were, in other parts of town, wholly subsidized or otherwise covered by others. Reuse thus was in most cases considered a luxury—akin to paying more for an item than was necessary, usually to achieve what are considered intangible qualities. Developers are in the business of increasing capital; therefore, many of them would not (in developers' parlance) "drive a Cadillac when a Chevy will do." And they stayed away from adaptive use.

The economic situation, like the preservation movement, has changed dramatically. In fact, the economic prejudices against adaptive use have nearly done an about-face. High interest rates have made capital very expensive. High energy costs have discouraged people from making long or frequent auto trips for shopping or work; location also is now an important real estate criterion as far as transportation is concerned. Public subsidies of strip development have been deterred for a number of reasons: the inflated cost of new road and utility construction; the reluctance of towns to incur further financial obligations through bonds, etc.; the federal government's Interagency Coordinating Council, which was established to mediate local controversies over federally funded projects posing an adverse impact on a town or city.

Quite frankly, the argument of embodied energy means little in economic terms to the developer. The value of embodied energy is already reflected in the cost of a structure. Within the context of an individual building the value of embodied energy is grossly outweighed by other economic considerations. For example, a property that may receive a zoning variance allowing a larger structure, and thus a higher return on investment, will most

Bradbury Building (1893, George C. Wyman), Los Angeles. Despite its natural light-filled atrium and other landmark qualities, the building has been threatened repeatedly because of difficulties in complying with local building codes, including the suggested enclosing of the stairwell shafts. (Photo: Carleton Knight III, National Trust)

Old Post Office (1873-84, Alfred B. Mullett), St. Louis. Occupying an entire block and embodying some 3.6 million gallons of gasoline, the landmark was saved after a protracted effort for historical and economic reasons more than for its energy investment. (Photo: Library of Congress)

likely be demolished for new construction. The cost of embodied energy was part of the purchase price (depreciated over time), but it is still worthwhile for the owner to demolish and build anew.

Shaping Public Sentiment

The economic argument for embodied energy, therefore, is weak in individual cases. It is, however, a social issue with important implications for national energy consumption.

How, then, can the concept of embodied energy be used in a local controversy for the benefit of preservation? My biggest concern is that it be used intelligently, where it will have the most effect and credibility. A local preservationist ranting about embodied energy to the developer of a property will get nowhere. In fact, such a demonstration will reveal the preservation advocate's naivety about the accountability and responsibilities involved in private development (not something new).

The embodied energy argument is best used in shaping public sentiment. Any project that has reached the level of being a controversy will invariably involve the local government for various approvals, clearances, assistance with grants, etc. If that were not the case, the property owners could do what they wanted without any delay. This is where public sentiment is important, especially in a small community.

Although not always vocal or visible, public sentiment is tested and measured by elected and appointed officials before they do anything. There are many gambits to use in arguing the public and private benefits of rehabilitation. The important point to understand is that one must play to the audience. Preservationists should talk to the public about the value of conserving embodied energy through rehabilitation. They should use as well the arguments of higher property tax returns through recycling, utilization of existing public services, the higher proportion of rehabilitation expenses going to labor and other arguments with which we as preservationists are familiar. The property owner can be informed about lower construction costs, tax incentives, shorter construction time, etc.

In using the embodied energy concept, one must understand its weaknesses as well as its strengths. One must understand to whom the point is best addressed and understand that the argument is only one of many reasons for preservation—none of which, by itself, would swing the balance in a local controversy. In the case of a historic sun-dried adobe structure to be demolished and replaced by a contemporary sun-dried adobe structure, the embodied energy argument would not be one's strongest suit.

Detail of columns lining entrance to the Old Post Office, St. Louis. (Photo: Robert Pettus)

3. Alternative Energy Sources for Old Buildings

Demonstrating Renewable Energy Technologies

Steve Mooney and Web Otis

Renewable energy sources will be providing a major part of our nation's energy diet in the coming decades. American ingenuity is developing and refining technologies to harness energy from the sun, the wind and the tides. Conservation is proving to be our best energy investment.

The quest for a sustainable energy future has a new focal point in the San Francisco Bay Area—the Golden Gate Energy Center, located at Fort Cronkhite in the Marin Headlands of the Golden Gate National Recreation Area.

Here on the Pacific coast, a short distance from San Francisco, government, industry and concerned citizens are working to convert a former military site to an innovation and demonstration center for conservation and renewable energy technologies. The center will operate educational programs for the visiting public and will provide workshop, lab, conference and specialized support facilities for community groups, businesses and government agencies.

One of several military installations in the Headlands, Fort Cronkhite was constructed in 1941. Over the years, 12 of the original barracks and several other buildings were removed. In 1972 the complex was turned over to the National Park Service. Fort Cronkhite now comprises some 40 structures and more than 25 fortifications and military installations ranging from World War II 16-inch gun emplacements to a deactivated NIKE missile base. Most of the fort's facilities are sited on 30 of its total land area of 400 acres. In 1980 Fort Cronkhite's buildings were listed in the National Register of Historic Places.

Opposite: Golden Gate Energy Center, located in the former Fort Cronkhite in the Marin Headlands of San Francisco Bay. (Photo: Golden Gate Energy Center)

At the instigation of the National Park Service, the U.S. Department of Energy collaborated with local energy conservation advocates in late 1978 to address the Park Service's dual concerns of saving buildings and saving energy. The result was a plan for the Golden Gate Energy Center. The center would capitalize on the large amounts of energy invested in the construction of 12 of the Fort Cronkhite buildings by renovating them for use as a center where local environmental groups, energy organizations, energy-related businesses and other professionals would promote the use of conservation and renewable energy technologies. Not only would the center reuse the embodied energy of the buildings, it also would weatherize the buildings to conserve energy and install renewable energy technologies to capture energy. All modifications would be designed according to National Park Service historical guidelines.

The Golden Gate Energy Center was inaugurated in May 1980. It is a private, nonprofit organization guided by a board of directors and several advisory committees drawn from both public and private groups. Funding is provided by private foundations, industry and government agencies, including the Department of the Interior and Department of Energy. The center operates educational programs for selected audiences, and it plans to offer programs for the visiting public (more than one million visitors to the Marin Headlands annually). The center also will provide workshops, laboratories, conference facilities and other support facilities for community groups, businesses and government agencies.

Solar-Conscious Siting

The National Park Service designated 12 buildings within a cantonment of 22 bar-

World War II barracks being converted into educational facilities for the Golden Gate Energy Center (Stoller/Partners). (All drawings: Courtesy Golden Gate Energy Center)

racks for use by the center. The siting of the cluster exhibits a high degree of inherent solar-conscious design. The buildings are grouped in three rows on seven acres of land that gently slopes toward the north, gaining approximately 10 feet in elevation for each 60 horizontal feet. The siting of the rows is such that buildings in the south row do not shade buildings in the north. The rows also are oriented on an east-west axis with the long axes of the buildings perpendicular to the south, providing unobstructed solar access.

Seven of the center's 12 buildings are two-story barracks. The structures are identical, each measuring 80 by 30 feet. The red composition shingle roofs are gabled and provide summer shading for the second-story windows. A skirt roof is located between the two stories, providing shading for the first-floor windows. The buildings are wood framed with wood siding. All windows are double hung, eight panes over eight. The foundations are concrete piers with a slab concrete foundation under the latrine.

Two of the center's buildings are former administrative quarters that are single-story, gable-roofed, wood-framed buildings measuring 51 by 25 feet. A former storage building is 45 by 25 feet, and a mess hall measures 93 by 25 feet. Once used as the fort's recreation hall, a large one-story building measuring 120 by 40 feet is located behind the uppermost row of structures. The center will be reusing the large quantity of energy that was invested in the construction of these buildings, plus the streets, sewers, storm drains and other utilities to which they are connected.

Administrative offices and conference facilities are currently located in one of

Proposal for conversion of one barracks into the main reception and exhibit center.

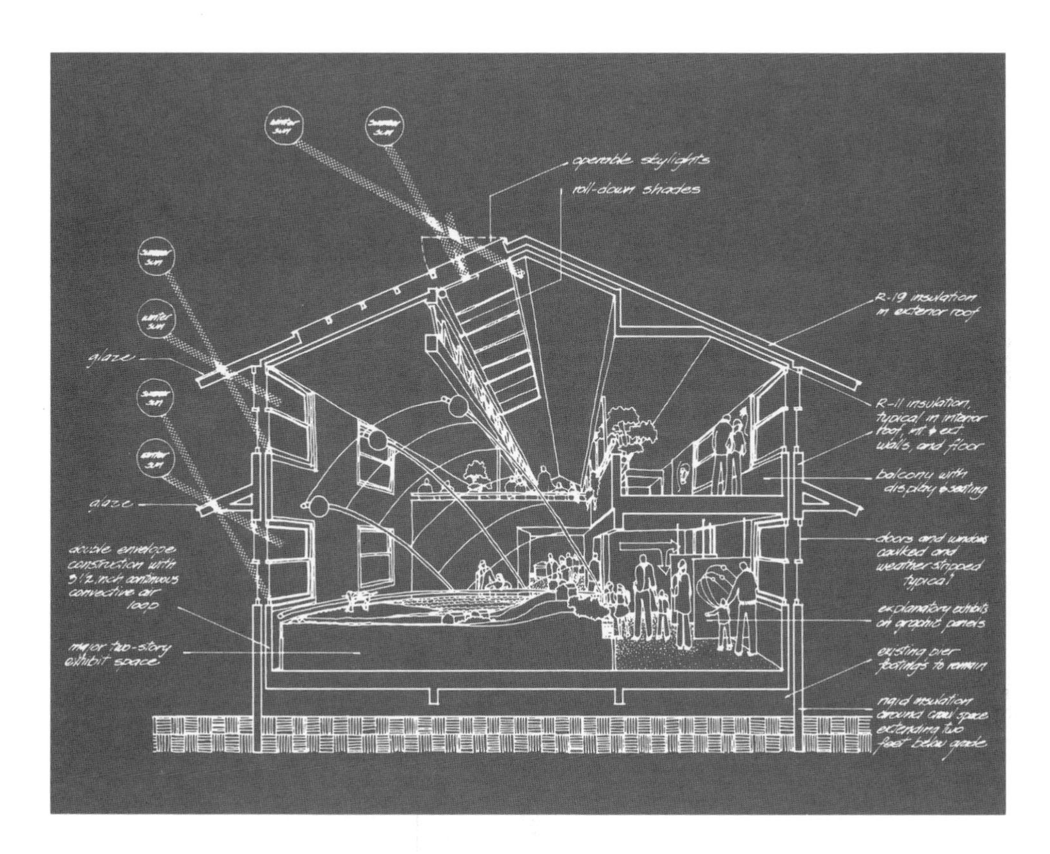

the barracks in the south row. The remaining two buildings in the south row will be used for client organization office space. Three buildings in the middle row will be converted to exhibit centers. A resource center also will be located in the middle row. The storage building in the uppermost row will be used as a building performance test model. A multiuse workshop is designed for the mess hall, and the two military administrative quarters will be used for small seminars and conferences. The recreation hall will be remodeled as the main conference facility, accommodating up to 300 participants.

Enhancing Fortuitous Design

The Golden Gate Energy Center and the National Park Service agreed on basic design criteria that have served as guidelines for the preparation of preliminary architectural plans. These criteria govern the preservation and adaptive use of the structures in accordance with the Secretary of the Interior's Standards for Rehabilitation and Guidelines for Rehabilitating Historic Structures. Final plans for individual buildings will be submitted for approval as they are completed.

Renovation of the Fort Cronkhite buildings by the center is directed toward enhancing the good climatic design fortuitously exhibited in the original construction. Major modifications involve energy conservation and the application of renewable energy technologies.

The buildings are without interior walls. Insulating will be relatively easy and will not result in change to the exterior appearance of the buildings. Proper vapor barriers and venting will be established to prevent undesirable condensation of moisture in the walls. Interior plywood

Plan for exhibit space and supporting work space with an attic solar collector.

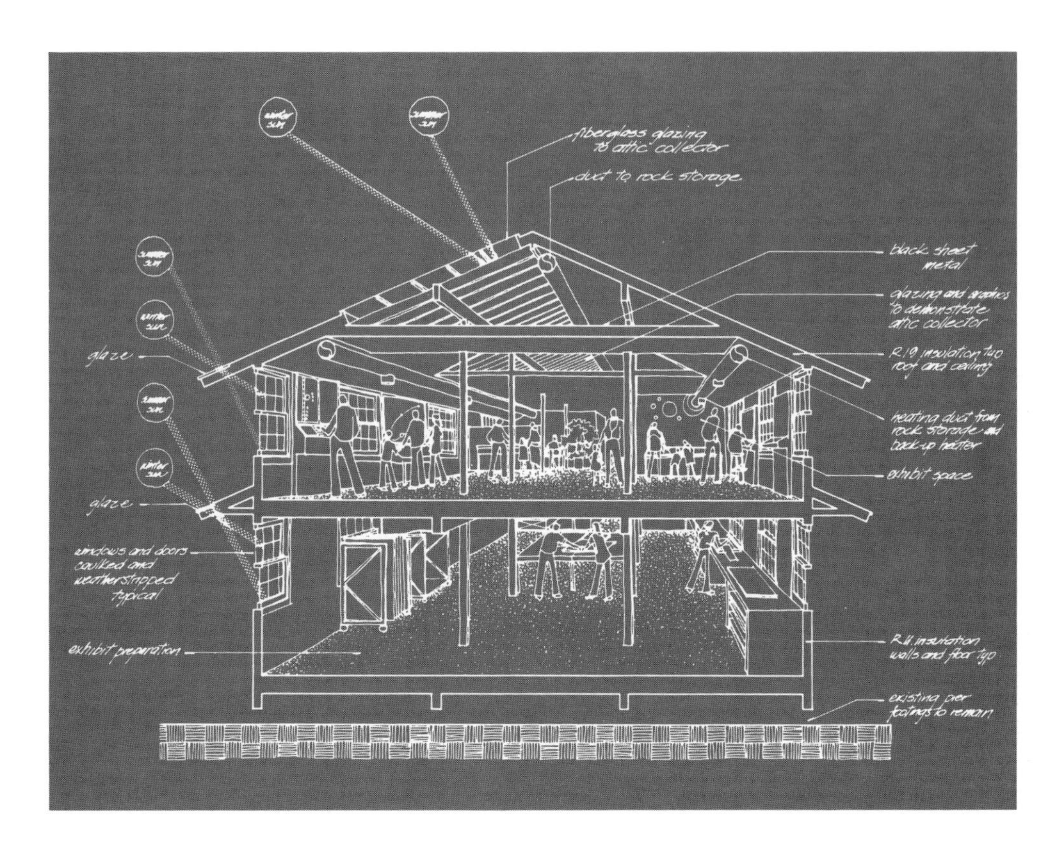

sheeting also will be installed to satisfy earthquake protection codes. In addition, all buildings will be weatherstripped and caulked.

Solar heating panels and skylights will be installed on buildings in the middle of the complex and will be mounted flush with existing roof lines to minimize visual impact. The solar panels will heat water for day and visitor use or heat air for supplemental space heating. Skylights will provide natural lighting and a source of heat. To increase direct solar gain and contribute to space heating, glazing will be increased on the south side of buildings located toward the interior of the cluster. Additional window space will incorporate the line of the existing skirt roofs and the rectilinear shape of the buildings.

Connections between buildings in the middle row (the exhibit buildings and resource center) will be constructed to facilitate access for the handicapped and provide a greenhouse environment for the cultivation of plants.

The basic objective of the design criteria is to preserve the historical character of the buildings and to minimize irreversible changes. To reduce the visual impact of renovations, exterior building modifications will be applied only to buildings on the interior of the cluster. In addition, the existing color, roofing style and siding details will be maintained in all improvements to building exteriors.

Demonstration of Options

The renovations also will be used in the center's demonstration program. Weatherization of the buildings will demonstrate a variety of choices for insulation, weatherstripping, window coverings, natural and artificial lighting methods and other

Proposed media resource center features a skylight.

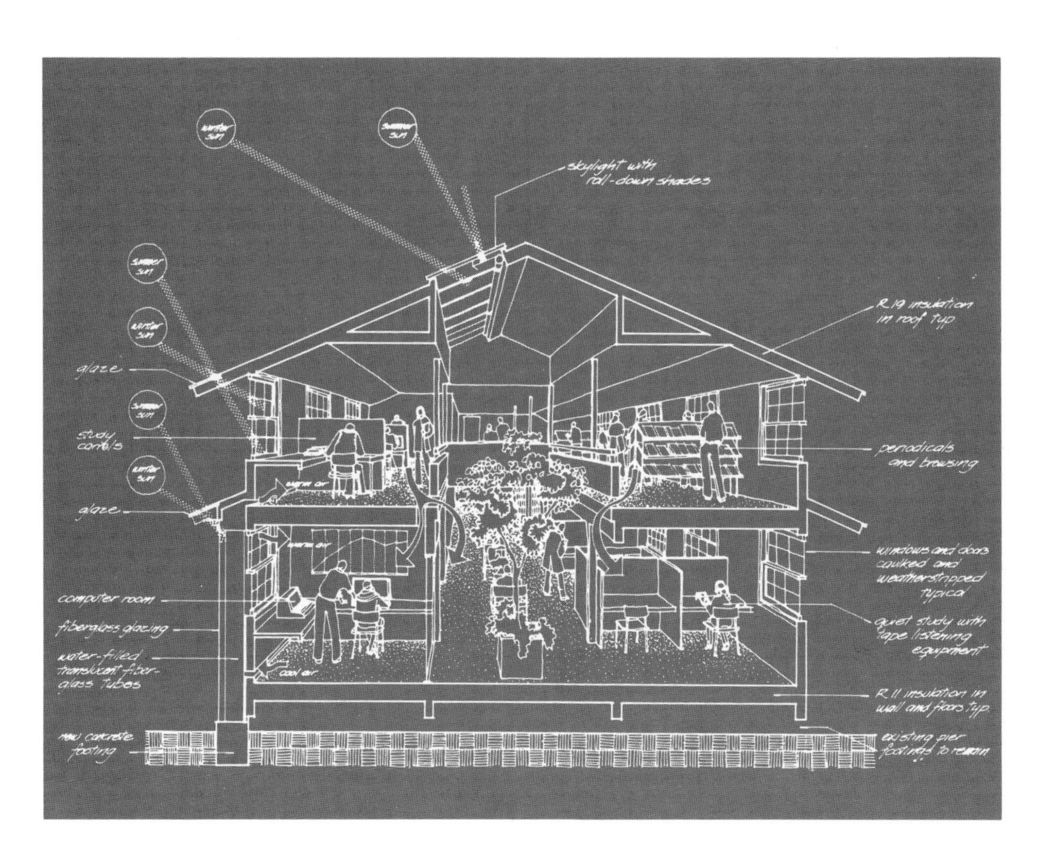

conservation improvements.

A controlled testing situation will be established in which several uniform buildings will be extensively weatherized in different configurations and then monitored for subsequent performance. An unimproved building will serve as a control to which the heating requirements of the improved buildings will be compared. This will provide an opportunity for quantitative analysis of the thermal performance of weatherized versus unweatherized structures.

Because seven of the buildings are identical and aligned in two rows with east-west orientation, they offer an unusual resource for the comparative testing and analysis of retrofit solar heating and cooling methods. With excellent exposure to the sun, these experimental shells can be used to demonstrate and test the comparative performance of various solar energy collection, storage and distribution systems. Each of three or four buildings will be retrofitted with a different design. Monitoring equipment will compare the relative performance of each system.

Renovation of the center's administrative office building was completed in October 1980. The upper floor of the former barracks has been converted to 2,400 square feet of attractive, energy-efficient office space. The ceilings are insulated to R-30 and the walls to R-11. Window frames are foam caulked, and windows and doors are weatherstripped. Windows and baseboards are trimmed in 1-by-4 redwood. Six T-partitions provide 12 workspaces. Pacific Gas and Electric, the local utility company, provided an energy-efficient lighting system. The bottom floor is used as a conference facility. It is renovated with the same design as the

Workshop plan features a skylight with reflective louvers and thermal storage in the ceiling tiles.

Product exhibit area and related work space includes wall solar collectors.

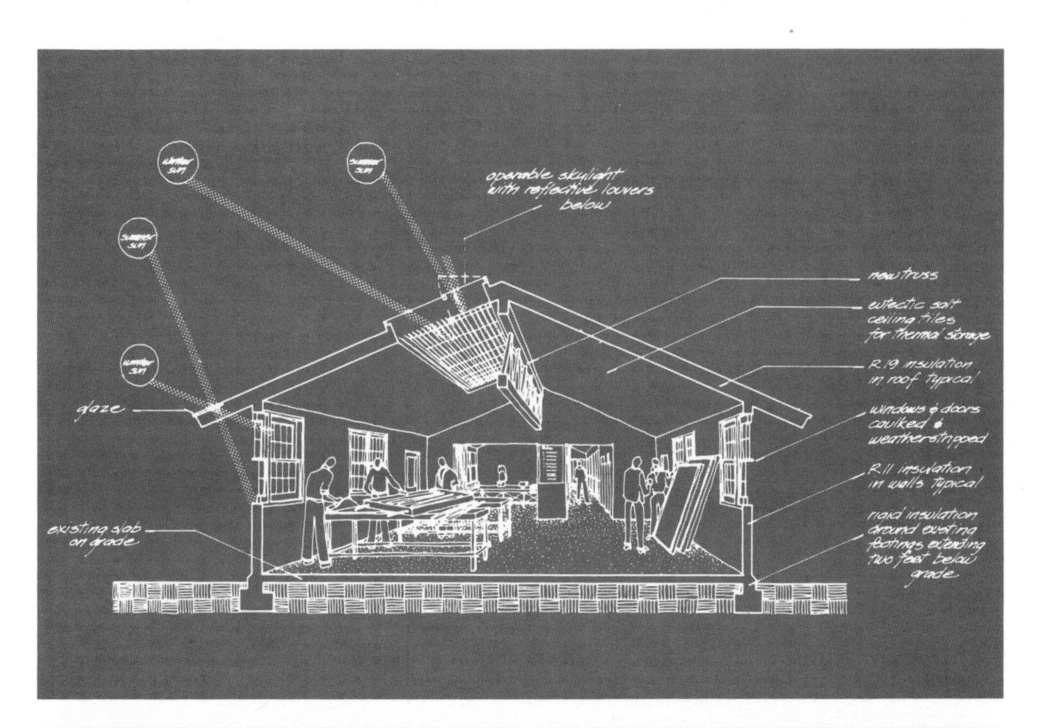

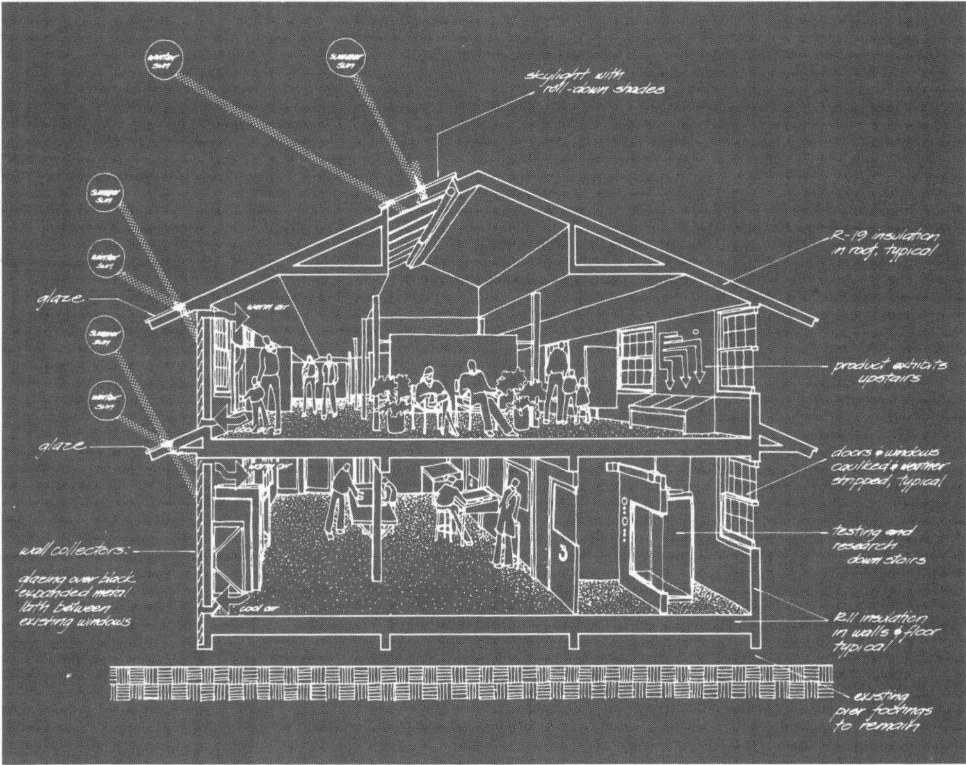

Wind energy conversion tests are being conducted on the Fort Cronkhite site as part of the center's programs.

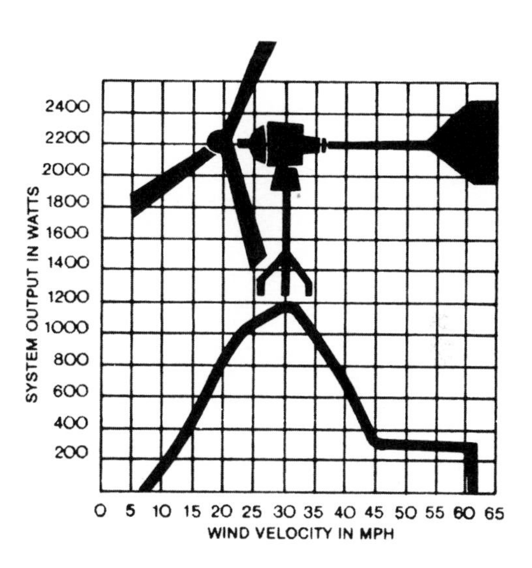

upper, taking advantage of an open floor plan to provide comfortable accommodations for up to 100 participants.

The program development staff anticipates completion of a second office building by February 1982, the seminar facility by April 1982, the workshop by May 1982, the main conference center by November 1982, the first laboratory-exhibit building by September 1982, the second exhibit building by April 1983, the third exhibit building by September 1983, a third office building by March 1984, the second seminar building by January 1984 and the resource center by November 1984.

The center's preservation and reuse of these buildings will save the large quantity of energy that was invested in the construction of the facility in 1941. Through energy conservation and application of renewable energy technologies, day-to-day operational energy also will be reduced considerably. Saving buildings saves energy, and given the uncertain energy future of the next 20 years, agencies that have responsibility for the preservation of buildings can take advantage of the opportunity to set a visible example for energy self-sufficiency. It is an opportunity for a no-risk investment with a high return.

Retrofitting with Passive Solar Energy

Fredric L. Quivik

Preservationists now talk a great deal about old buildings being inherently more energy efficient than many new ones and about the value of preserving energy embodied in old buildings. But these arguments go only so far. Old buildings that conserve energy are still less efficient than some newer buildings that incorporate contemporary conservation technology, and embodied energy interests primarily the policy maker, not the developer or occupant.

Consequently, it is time for preservationists to move forward and consider ways to make old buildings slated for preservation and rehabilitation even more energy efficient. Energy-conserving retrofit techniques, such as passive solar applications, should be pursued to enhance the value and utility of old and historic buildings. Such a pursuit would lead also to more acceptable design standards for rehabilitation projects in general. Seventy percent of the buildings that will be in use 30 years from now are already in existence, so there is a great need for the preservation movement to develop retrofit design standards that will maintain the character of those existing buildings.

Before discussing specific retrofit techniques, two building types should be distinguished: small and large. Small buildings have skin-dominated energy consumption; that is, they have relatively small volume-to-surface-area ratios. Thus, most of the energy consumed in small buildings is related to heat gained or lost through the building envelope (skin) rather than to mechanical systems (heating, cooling, air handling, electrical, conveyance, etc.) within the building. Large buildings, on the other hand, have large volume-to-surface-area ratios, and their energy consumption is systems-dominated. The remarks here are limited to small buildings.

Opposite: A sun space was created through the addition of a glass wall (1978, Anderson Notter Finegold) added to Mechanics Hall (1855-57, Elbridge Boyden), Worcester, Mass. (Photo: © Steve Rosenthal)

By far the most cost-effective action to make a building more energy efficient is weatherization. Therefore, the solar retrofit techniques described here should always follow comprehensive insulation, caulking, weatherstripping, storm window installation, etc. Mechanical system adjustments take precedence as well—the furnace tuned efficiently and a night setback installed on the thermostat. After all conservation measures have been fully explored and implemented, options for passive solar retrofitting should be studied.

Four basic passive design concepts are used in new construction and are relevant also to retrofit projects:

Direct gain. The most straightforward passive design concept is direct gain, in which solar energy penetrates south-facing windows, warms the interior space and is stored in water containers, masonry floors or in the thermal mass of the structure itself.

A direct-gain retrofit of a historic building generally entails optimizing the windows. In a cold climate, two or possibly three layers of glass should be installed. Movable or nighttime insulation also should be considered. These steps make south-facing windows net heat gainers rather than net heat losers. The windows facing east, west and north also must be treated. West-facing windows will gain a tremendous amount of undesirable heat on summer afternoons, so shading should be part of the retrofit. East-facing windows may help warm interior space in the morning, but after that they tend to lose heat; they should be equipped with multiple glazing and preferably movable insulation to conserve heat after the sun passes. North-facing windows always lose heat; an adequate number for daylighting needs should be maintained and the rest closed off in a manner compatible with the facade.

Mass wall. A mass wall passive design is one in which the solar energy passes through a layer or layers of glazing and immediately strikes a massive wall. The converted solar heat is stored in the wall and slowly moves through it to the living space. The phenomenon of convection also may be used in this system. Air flows through vents in the bottom of the wall, is warmed by the sun in the space between the wall and glazing and rises to the top of the wall, where it flows back into the living space through upper vents.

The native peoples of the southwestern United States used this basic design centuries ago. They built massive pueblos, using the walls to collect solar energy during the day when it was not needed; in the evening, the heat stored in the walls would warm the living spaces.

Buildings with masonry walls facing south may lend themselves to mass wall retrofitting. One need only attach a layer or two of glazing to an existing brick wall and the passive system is in place. Existing windows can be used to vent air to and from the living space.

Sun space. An attached sun space is essentially a greenhouse attached to the south side of a building. Air in the greenhouse is heated by the sun and vented into the main living space. Thomas Jefferson used this technique at Monticello.

People all over the country are attaching greenhouses to their homes. Besides the heating benefits, greenhouses provide a space to raise vegetables and an opportunity to prolong the garden season. Excellent publications are available on how to design, build and use greenhouses.

Microload passive. This concept, sometimes called superinsulation, is the latest development in energy-efficient design. Until a few years ago, thoughts of solar-heated buildings always included solar collectors, pumps, fans, remote storage, etc. Then it was realized that buildings themselves could be designed more efficiently and did not necessarily need complex active systems, so the three basic passive designs just described were refined. Now people are starting to use tremendous amounts of insulation in buildings that are also extremely airtight. In these superinsulated buildings, the heating load is reduced to perhaps a tenth of what similarly sized residential buildings would require.

In Regina, Saskatchewan, where homeowners normally pay about $1,000 a year for heat, microload passive homeowners are paying about $100 a year. Most of the heat required in these houses comes from appliances, body heat from the residents, lights and a few south-facing windows. Some of the microload passive houses built in the United States are so energy efficient that no furnace is installed; the hot-water heat or small lengths of electric baseboard are used for backup heat. In these cases, the additional $2,000 or $3,000 spent in insulation is offset by the money saved by not installing a furnace system.

Infiltration rates have been reduced drastically in these houses. Typically, a residence measures one complete air change every hour. In some superinsulated houses, it takes 20 hours for one complete air change. When such a small amount of air is being exchanged, a variety of problems can develop: Humidity may become troublesome; microorganisms may build up and cause ill health; radioactive isotopes may escape from radon gas in the building materials and pose a health threat. However, these problems are managed easily by installing an air-to-air heat exchanger. These units bring fresh air into the house and exhaust old air while transferring 80 to 90 percent of the heat available in the exhaust to the incoming air. Health hazards thus are eliminated without losing heat.

Four types of passive solar energy systems: direct gain (top left), Trombe wall (top right), water wall (bottom left) and solar greenhouse (bottom right). (Drawings: U.S. Department of Housing and Urban Development)

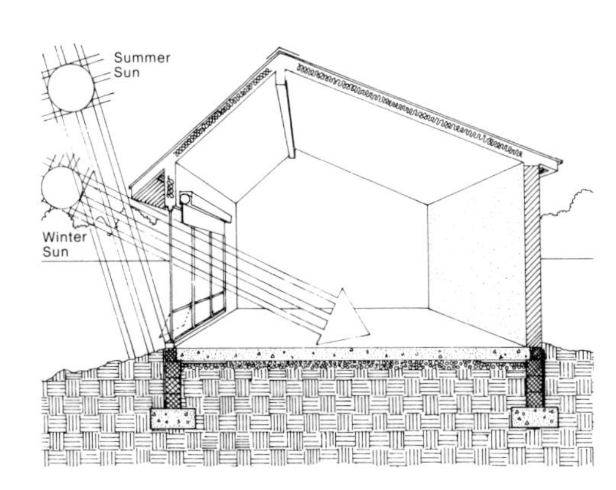

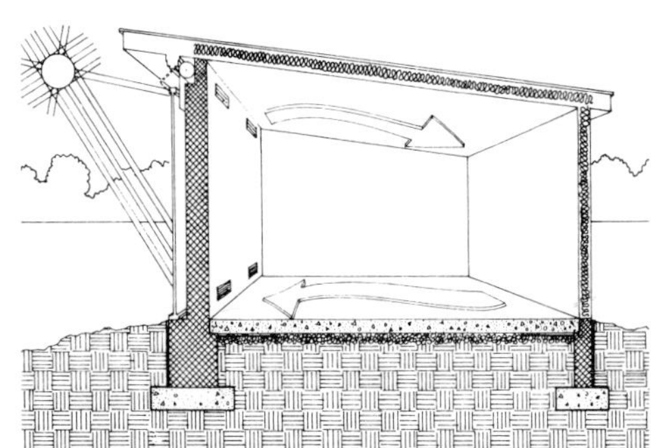

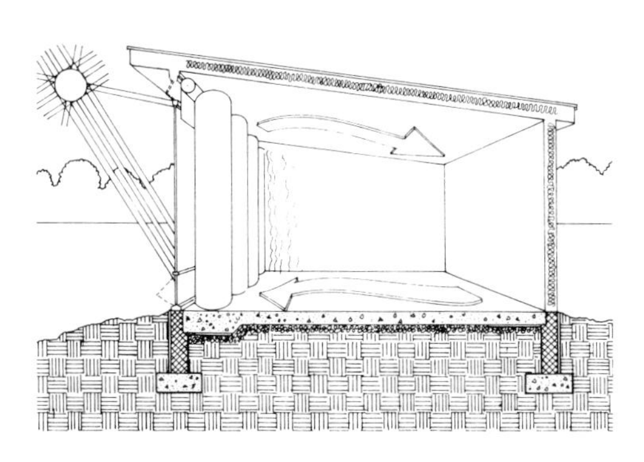

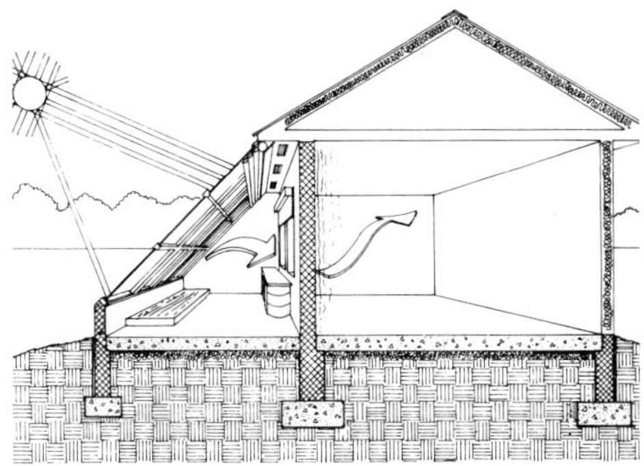

Hamill House (1867), Georgetown, Colo., with a solarium added in 1879. (Photo: Carleton Knight III, National Trust)

Kalwall Corporation warehouse, Manchester, N.H., undergoing retrofitting (top) and with thermal wall in place (bottom). (Photos: Kalwall Corporation)

Little is known about the applicability of the microload passive design concept to existing buildings. Many more variables are associated with existing buildings than with new construction. However, experience to date indicates that, in some cases, a superinsulated retrofit will be more cost effective than passive solar retrofit applications. The National Center for Appropriate Technology plans to perform a superinsulated retrofit during 1981 and will be able later to report more fully on the subject. Until then, at least, the implications for the preservation movement are promising: A superinsulation retrofit may demand much less change to the exterior character of a building than does a solar retrofit.

There are two major reasons to retrofit: to save energy and to save money. Two examples, one in New Hampshire and the other in Montana, illustrate these savings.

In Manchester, N.H., a brick warehouse built in the 1920s and owned by the Kalwall Corporation, a leading manufacturer of solar products, was retrofitted with two layers of the company's glazing material. About 2,000 square feet of the south wall were glazed as a product test and as a means of advocating another use for the material. The retrofit, if built by someone other than Kalwall, would cost between $10 and $12 a square foot, depending on labor costs. Designers projected that the retrofit would reduce the heating demand by around 140,000 Btu's per square foot of retrofit per year. Preliminary results have shown that the retrofit has saved 2,000 gallons of fuel oil per year, or about a gallon per square foot of retrofit. Assuming 90 cents per gallon for heating oil (1979-80 costs) and a 60-percent boiler efficiency, the wall will pay back in 11 to 13 years.

In Butte, Mont., a U.S. Department of Energy grant to the Butte Indian Alcohol Program Halfway House funded a 600-square-foot retrofit. The National Center for Appropriate Technology provided

Butte Indian Alcohol Program Halfway House, Butte, Mont., before retrofitting (top) and following installation of thermal wall. (Photos: Fredric L. Quivik)

technical assistance during construction in summer 1980 and is monitoring the performance of the building. A computer model developed at NCAT for predicting the performance of both newly constructed and retrofitted buildings with passive systems indicates that the halfway house retrofit should produce about 59,000 Btu's per square foot per year. Before being retrofitted, the wall lost about 24,000 Btu's per square foot per year. (Steady-state heat loss calculations show about 47,000 Btu's per square foot per year lost through the wall, but experience indicates that for south-facing walls the actual loss is about half that.) Thus, the net benefit of the retrofit will be 83,000 Btu's per square foot per year. Using the 1979-80 winter fuel costs for Butte and a $10-per-square-foot construction cost as a basis for computation, the work should have a 25-year payback. The difference in the payback periods in the Butte and Manchester examples is a function of the difference in fuel costs at those two localities.

The effect of a solar retrofit on existing building materials is a concern. In the case of the halfway house, it is suspected that the temperature range in the south-facing brick wall will be smaller after being retrofitted. The range before retrofitting was between −40°F in the worst winter conditions and 160°F on the hottest summer days. After retrofitting, the high temperature probably will not change, but the wall temperature rarely will fall below the freezing point. Experience at the Kalwall warehouse has supported an additional supposition that the elimination of rain and wind striking the south wall also will provide a net benefit to the building materials.

Impact on Historical Character

The most critical concern in the eyes of preservationists is the impact a retrofit will have on the appearance and character of a historic building. There has been an

Faneuil Hall Marketplace (1976, F.A. Stahl and Associates; Benjamin Thompson and Associates), Boston, with the new glass-canopied market stalls in place on the old Quincy Market (1826, Alexander Parris). (Drawing: The Rouse Company)

apparent conflict between traditional preservation values and the social value of conserving energy. Yet, that is exactly why preservationists should get involved in retrofitting: to help resolve the conflict.

In the past, preservationists have encountered other conflicts between preservation and different social goals, and the record reveals precedent-setting solutions to the conflicts. On the issue of handicapped access, for example, preservationists generally have realized that it is important to compromise traditional design values to secure access to historic buildings for handicapped persons. Preservationists helped resolve this conflict by developing design solutions that were appropriate to both social goals.

Another conflict involves the need to make old buildings viable in today's economy. Faneuil Hall Marketplace in Boston is an example of a successful design solution. Significant amounts of glass were added to the building to create seemingly outdoor market space protected from the elements, thus making the project economically feasible. Yet, the addition of the glass altered the overall shape of the building. What once was a long, narrow, multistory building now appears to be a long, narrow building with side aisles.

Another example is the Tivoli Brewery

Original proposal for the conversion of Tivoli Brewery (1890-91, main building), Denver, into a shopping complex with a glass-covered courtyard.

Sunapee Mill (1843-46), Claremont, N.H., as it looks today (top) and as it would appear with a Trombe wall to provide solar-heated water (bottom), a proposal found to be inappropriate. (Photo: Fredric L. Quivik. Drawing: P. Scharf. Historic American Engineering Record)

project in Denver, which was approved for rehabilitation under the Tax Reform Act of 1976. The original proposal for this shopping center conversion included a large glass-covered courtyard that would join all the buildings. The character of this building complex would change significantly. What once was a large building and several ancillary buildings would appear to be a megastructure.

Preservationists should be applauded for accepting such design interventions and in supporting these projects to attain economic viability. But from an energy standpoint, it is interesting to note that Faneuil Hall Marketplace and the original Tivoli Brewery proposal make absolutely no sense. All that horizontal glass is destined to cause overheating in the summer and a tremendous heat loss in the winter.

If Faneuil Hall Marketplace and the Tivoli Brewery are deemed acceptable as preservation projects, we must reconsider how retrofit projects are evaluated. Two examples that can be used for comparison are the halfway house in Butte described earlier and a mill building retrofit designed—but never executed—by the Historic American Engineering Record in Claremont, N.H., during summer 1978. The Butte project was deemed to be non-

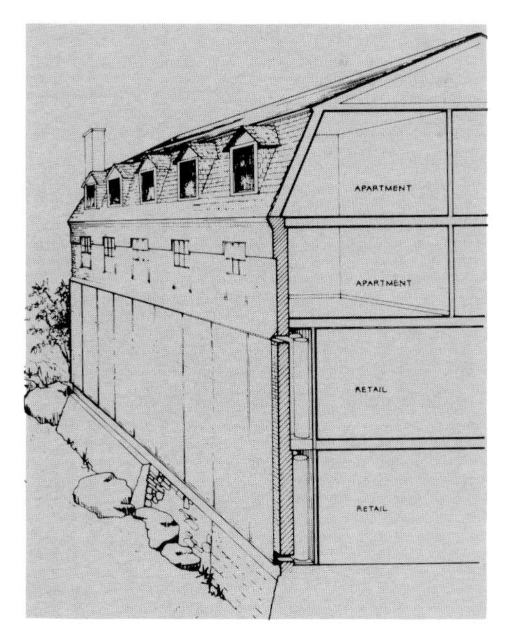

Monadnock Mills 2 (1853) and 6 (1910), Claremont, N.H., as they appear at present (top) and with a proposed glazed connecting arcade that would capture solar heat. (Drawings: M. Chrisney, Historic American Engineering Record)

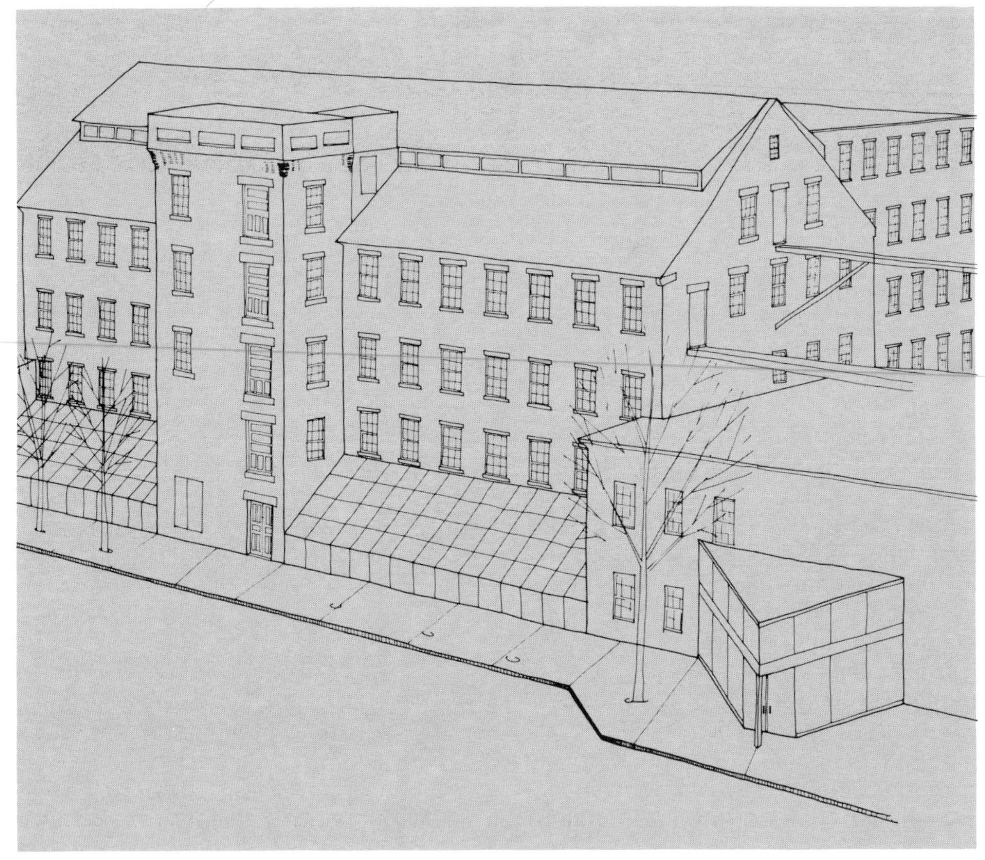

Site plan for a comprehensive rehabilitation of the Claremont, N.H., mill complex. (Drawing: M. Mook, Historic American Engineering Record)

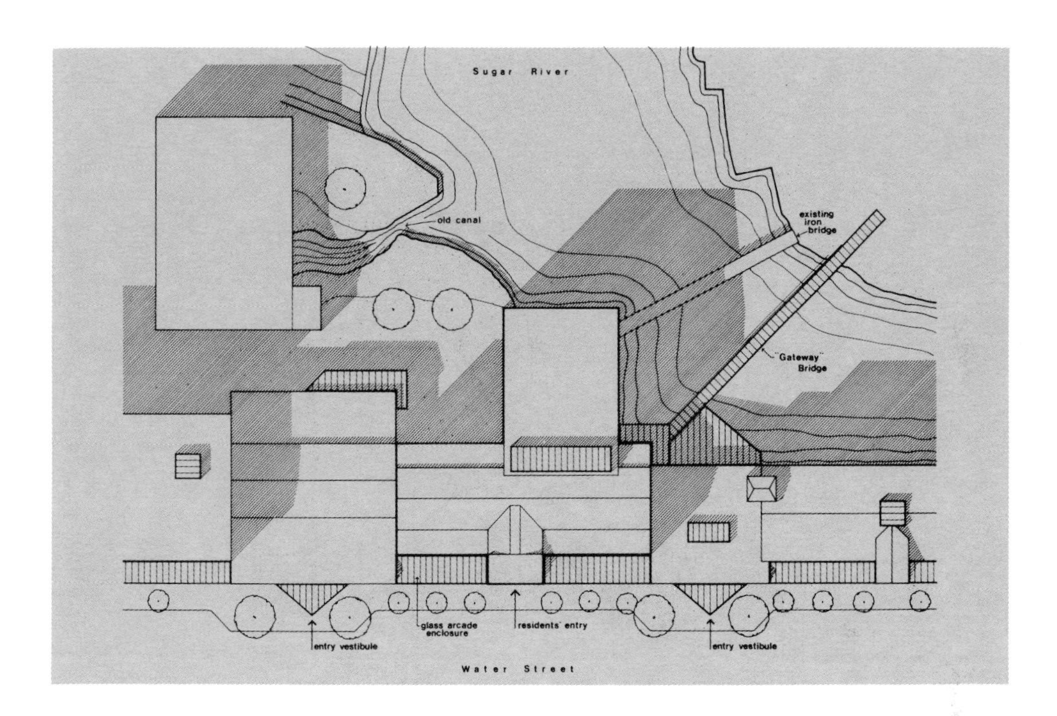

conforming to the Secretary of the Interior's Standards for Rehabilitation when submitted for Tax Act certification. The Claremont design, although never formally submitted for approval, was informally judged also to be inappropriate for the historic building.

The halfway house has indeed had its south facade altered by the retrofit, as the Claremont building would have, but both would remain freestanding buildings of the same shape and size as before with the same relationship to their districts and to their other three facades. And they are or would be more energy efficient. Meanwhile, the Faneuil Hall Marketplace and Tivoli Brewery projects alter the massing of those historic properties and probably worsen the energy consumption of the buildings.

Developing Flexibility

The historic preservation movement has not yet fully recognized the important role it can play in the national effort to conserve energy. Likewise, preservationists have failed to recognize fully the economic importance of energy conservation to the movement. The methodology to determine which historic properties can withstand energy-related intervention without adverse effects is not flexible enough. Instead of resisting such intervention, preservationists should become more actively involved in energy retrofit projects so that a more adequate methodology can be developed. Projects should be evaluated on a case-by-case basis. Energy-efficient retrofits that are compatible with the historical significance of older buildings can and should be developed.

The coin that rewards our efforts will have two sides: more historic and other old buildings saved because their operating costs will be reduced, and more people living in their existing housing or using their existing office, commercial, institutional and industrial buildings.

THE NATIONAL TRUST
HISTORIC PRESERVATION

Active Solar Applications in Old Buildings

Gary Long, AIA

An argument against energy conservation in old buildings and passive solar applications in new buildings is hard to find: we have rediscovered the ethos of resource husbandry and design with climate. But active solar collectors boldly in view on the roof of an old building are another matter. Active solar systems for old buildings are a difficult design problem, one that for architects and preservationists will become more pressing as the price of fuels goes up and the price of solar equipment comes down.

Some people still question the reality of the energy bind in which we find ourselves, but attitudes are indeed changing, and changing rapidly. Since the embargo of 1973-74, building owners who pay the fuel bills have no little sophistication in understanding energy and buildings. It behooves preservationists to obtain an equal sophistication in understanding building energy needs and the conservation and solar options available to meet those needs.

Three points in particular bear on the use of active solar systems in old buildings:

1. *Conservation comes before active solar use.* If the energy requirement of a building is cut in half, the collection system is cut in half.
2. *An active solar system is a mechanical system.* It should receive the same design consideration and treatment as, say, a roof-top air conditioner.
3. *Save the old heating system.* The goal of conservation of resources applies to old mechanical systems as well as old buildings. Make the new mesh with the old.

Opposite: Andrew Mellon Building (1915-17, Jules Henri de Sibour), Washington, D.C., the National Trust headquarters. An active solar collector is located behind the roof but is not visible. (Photo: Robert C. Lautman)

The problem is both a design problem and a preservation problem: how to place the hard-edged equipment of a new technology on an old roof without destroying the appearance of the building and how to integrate a new solar system with radiators, grilles and boilers that are themselves important elements of an old building.

It is critically important that an active solar system, like any other mechanical system, be considered only after all reasonable conservation measures have been taken. This cannot be emphasized enough. A reduction in a building's energy requirement, its energy load, results in a reduction in the size of the applied solar system. Measures related to existing systems for heating, ventilation and air conditioning and to existing boilers and refrigeration equipment should be considered as steps in the integration of new solar hardware with old buildings.

The listing of basic energy conservation steps includes the seemingly obvious as well as the technically arcane; some procedures can be effected by building owners and users, and some are the province of consultant architects and engineers. But in every case the point is that conservation measures must be identified before sensible solar design can take place.

Solar Collection System

The sun provides mechanical energy through wind machines, electrical energy through photovoltaic conversion and thermal energy through direct collection for space heating and domestic hot-water heating. Thermal energy from the sun also can be used for cooling through absorption or desiccant processes, but

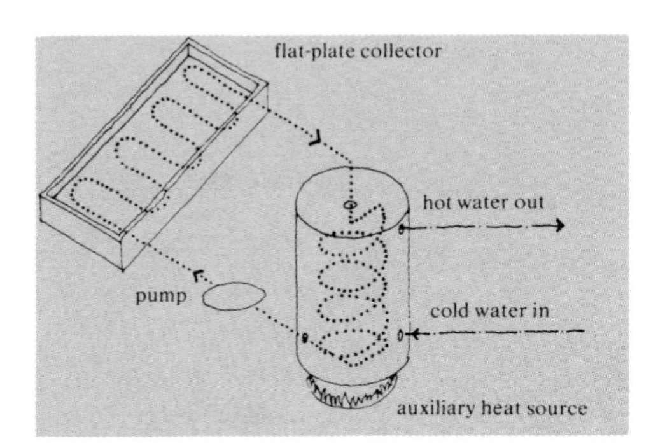

The basic solar hot-water heating system. (Drawing: Courtesy Gary Long)

this use is not of the same order of interest as space and water heating.

An active solar thermal collection system is composed of six elements:

1. *Collector*—the visual problem
2. *Storage*—the where-do-you-put-it problem
3. *Working fluid*—air, water or glycol-water solution
4. *Auxiliary*—the support system for long cloudy periods
5. *Control system*—to turn fans and pumps on and off
6. *Distribution system*—for putting energy where it is needed, the match-to-existing-equipment problem

The elements of a solar electrical collection system (photovoltaic system) are functionally the same as those of a solar thermal system. The collectors are visually similar, the storage is typically batteries and the auxiliary is perhaps a gasoline generator. The use of photovoltaic systems is at present limited by cost to building sites not served by public utilities, but within the decade they should be as readily available as thermal collection systems are now. The visual impact of the collectors is the same for both thermal and photovoltaic systems.

Wind machines for water pumping or electrical generation do not present the same design problem for preservationists as thermal systems. However, the need for these machines, except perhaps for isolated buildings, is not immediate, and their use is limited by access to good prevailing wind velocities. The wind energy collected varies with the cube of the wind velocity; that is, if a 10-mile-per-hour breeze produces 2 watts, a 20-mile-per-hour breeze produces 16 watts. For this reason, most wind generators make little economic sense at prevailing wind velocities less than 10 miles per hour, velocities uncommon to urban situations. If a wind tower is an unwarranted intru-

Solar space heating system integrated with existing equipment. (Drawing: Courtesy Gary Long)

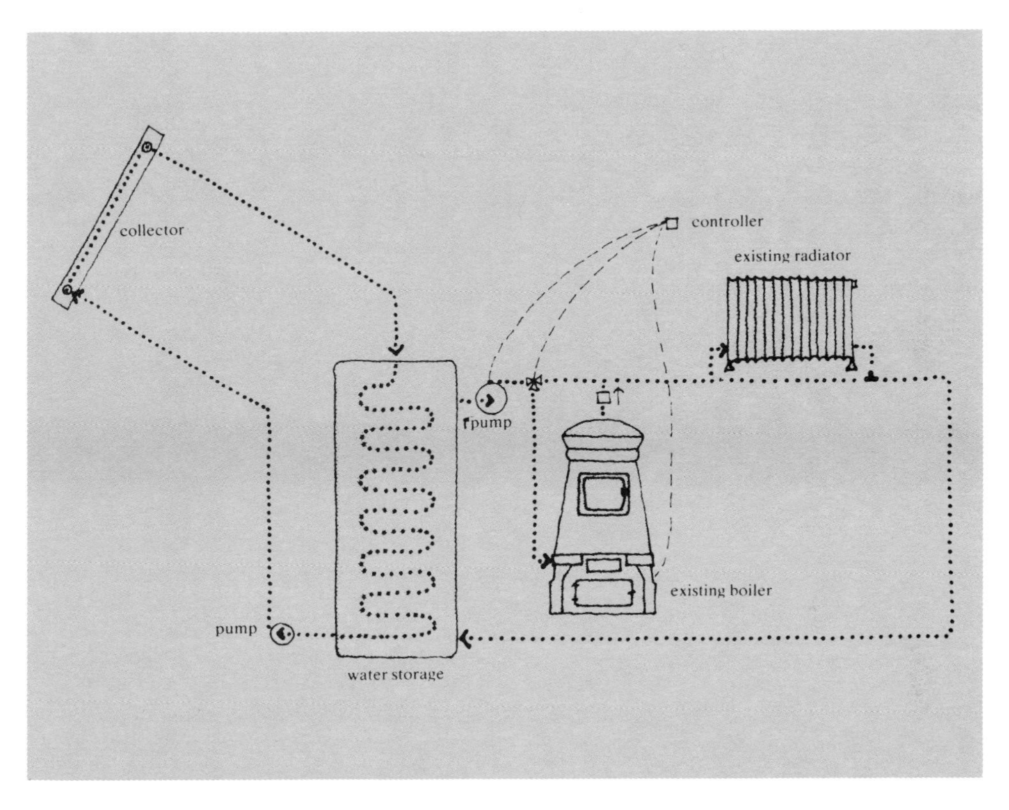

sion, it can easily be placed in a location remote from historic buildings.

The sizing of the collector and storage for any solar system is an optimization problem. To provide collection and storage for the worst extended cloudy period would be inordinately expensive. In practice, the percentage of the system requirement met by the solar equipment, the solar fraction, is always significantly less than 100 percent. Solar fractions of 30 to 70 percent for space heating or domestic hot-water heating are typical.

At times, the fit of an active solar collection system to an old building can produce an awkward clash between the aesthetics and technologies of different eras; at others, the fit can be a smooth integration of old and new values.

Solar design produces a visual problem of collectors at one end of the system and a problem of integration with existing distribution elements (grilles, radiators, etc.) at the other end. Collectors are an aesthetic problem, and integration with existing heating elements is a technical problem. In addition, the auxiliary system (which may already be in place), storage, controls, pipes, ducts, etc., have their own demands but generally follow the decisions made with respect to the collectors and the distribution system.

A collector works like a greenhouse: The sun streams through a glass or a plastic cover and is trapped. The cover is transparent to the sun's energy but is largely opaque to the long-wave radiation of the heated absorber plate. The energy collected on the absorber plate is carried away to storage through pipes or ducts.

Flat-plate collectors are now well known. They are big boxes, generally black metal with a shiny appearance. A collector system for domestic hot water for a typical house generally is some 50 to 80 square feet in size. The collector area

Pomona Valley, Calif., bungalow with a roof-top solar water heater in 1911. (Photo: From A Golden Thread *by Ken Butti and John Perlin)*

for significant heating of a small building might be a quarter to a half the building floor area.

Focusing collectors of more complex shape using reflectors or special lenses are available but are used to provide high fluid temperatures not generally a requirement of retrofit programs.

The large size of flat-plate collectors (on the order of three feet by seven feet) makes their use with smaller building elements of earlier construction periods difficult. If the appearance of the old building is important, a remote location of the collectors on a shed roof or in a yard area is appropriate.

If the existing building is flat roofed and protected by a parapet, there is no visual problem at all. Collector arrays can be oriented for maximum solar advantage and can be used with mechanisms that turn the collectors to follow the sun. This is the most desirable application for existing buildings.

If a decision is made to place collectors on an existing sloped roof, further questions of orientation need satisfaction. Although there is some latitude for placement, there are, nevertheless, certain economic limits. A collector that collects only half as much per unit area must be twice as big. Reasonable limits are 30° from south in direction and a tilt angle from the horizontal of latitude 15°.

Collectors tending to the vertical are useful in northern latitudes for space heating but collect little energy in the summer and should not be used for domestic hot-water heating. The optimum tilt for producing domestic hot water year-round should be close to a tilt angle from the horizontal equal to the latitude of the building.

If collectors cannot lay flat against an existing sloped roof, the visual problems are compounded tenfold with supports and exposed piping. If this is the case, consider abandoning the exercise—or look for a flat-roofed building!

If the building has no existing heating

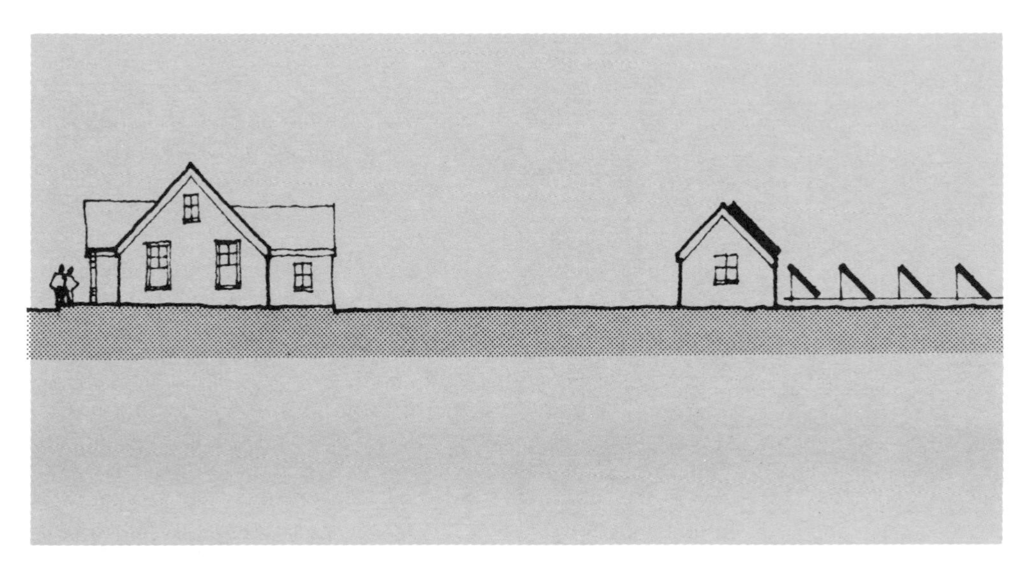

A remote array of solar collectors, designed to lessen the potential adverse design impact on a building. (Drawing: Courtesy Gary Long)

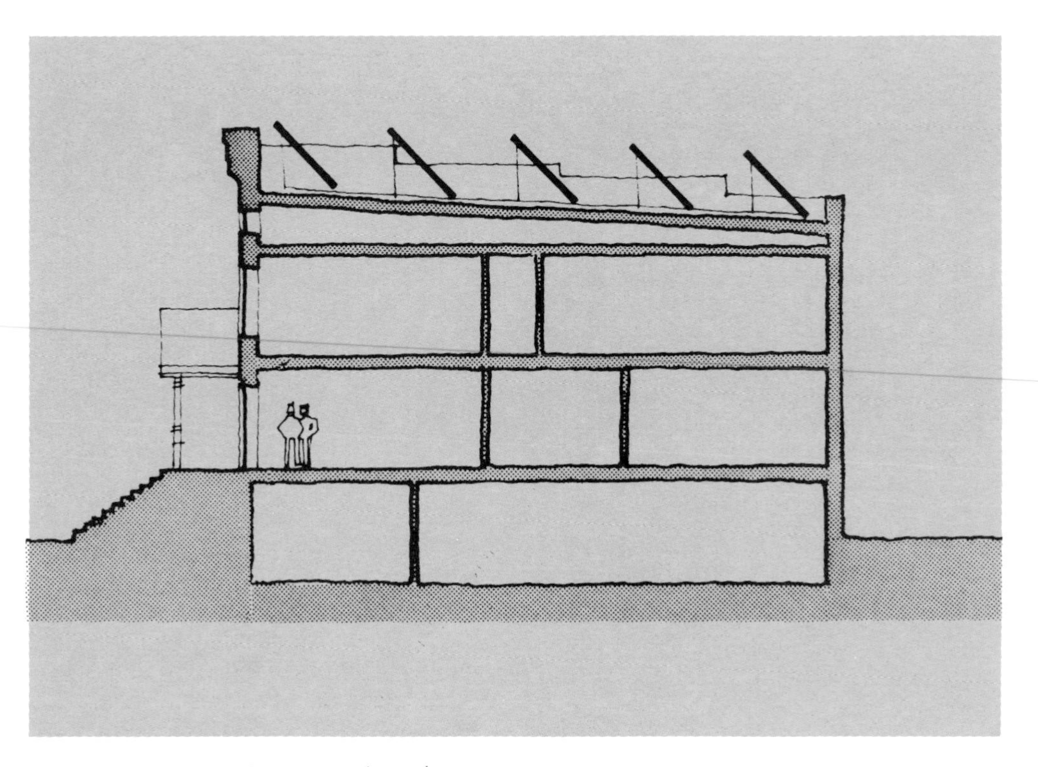

Collectors located behind parapet roofs, an alternative means of preventing a disruptive appearance. (Drawing: Courtesy Gary Long)

Raised gable on a historic house balances the solar collector, located on a roof section facing the back yard to avoid visibility from the street. (Drawing: From Residential Solar Design Review *by the American Planning Association)*

system, design procedures for active solar applications simply follow contemporary practice within constraints of the availability of space for equipment. For the building that does have an operating heating system, a bias is offered: Keep it. A radiator or an air register is as valuable a piece of the fabric of a building as a door or a window or the base molding. Beware of engineers or mechanical contractors who shake their heads and say without detailed justification, "That antiquated system has to come out." In many instances, the solar system can act as a preheater for an existing boiler or furnace, and the existing system can be left intact. One of the biggest problems observed in adaptive use or rehabilitation is the insensitive addition of forced air ducts in buildings served originally with gravity air, steam or water systems. The original systems worked for years; often they still work.

It is true that gravity water and air systems are designed to operate in the vicinity of 160°F, a fluid temperature difficult to obtain with flat-plate collectors without serious loss in solar efficiency. (Higher temperatures are achieved by slowing the flow rate through the collector; but the higher the collector temperature, the more energy is lost to the atmosphere before it can be carried off the collector to storage.) However, with conservation measures that decrease the building heating energy load, lower temperatures frequently can be used for the same degree of comfort.

The simplest solar thermal application and the one most cost effective, because of its year-round use, is solar domestic hot-water heating. The array of collectors is connected directly by pipe to a large storage tank, which feeds the existing hot-water heater or has within it an auxil-

Myrtle Gardens (c. 1909), Springfield, Mass., a four-story apartment rehabilitated with conservation and solar elements (1980, Steven Hale and Anderson Notter Finegold). Development drawings show the new clerestory skylights, active solar collectors and a greenhouse. (Drawings: Courtesy Anderson Notter Finegold)

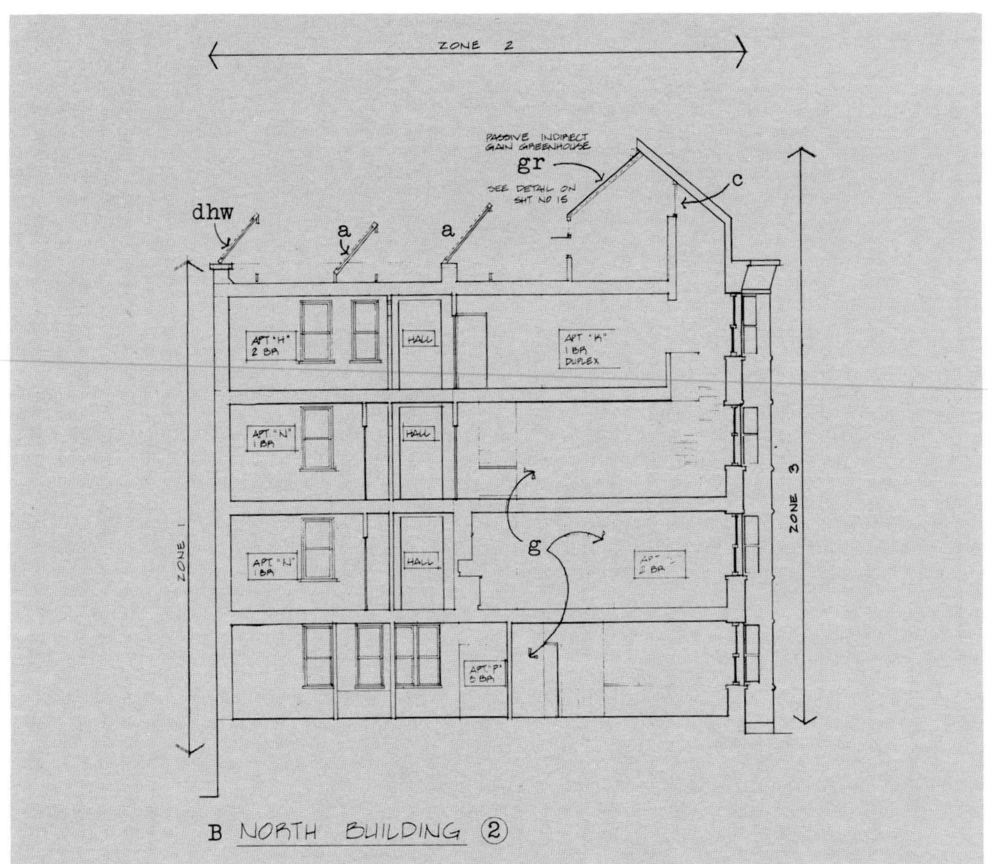

Sailors Snug Harbor Chapel (1855), Staten Island, N.Y. Installation of roof-top solar collectors resembling a historic standing-seam metal roof is being studied as part of the building's rehabilitation. (Photo: Sailors Snug Harbor)

iary heater that takes the place of the existing hot-water heater.

The direct application of solar systems to existing space heating systems is a bit more complex. Solar thermal storage generally is not a visual issue but certainly can be a technical problem. Three schemes are available: water tanks for liquid systems, rock bed storage for air systems and phase-change materials for either water or air systems. A two to five-gallon water tank or 100 pounds of rock bed per square foot of collector are typical requirements. The water tank and rock bed storage are simply a matter of heating the water or rocks for later use. The phase-change materials are different. As energy is added to the storage, the material melts but remains at a constant temperature (which is set by the nature of the material), just as ice remains at 32°F until the whole mixture is liquid. As energy is used from storage, the phase-change material gradually freezes, again remaining at a constant temperature until all the material is frozen.

The functional problem for storage is one of weight and volume, and these two questions must be addressed within the structure and space capacities of old buildings.

Active solar systems generally are not considered pretty when placed on old buildings. But they work, and when conservation and passive solar options are exhausted, active solar equipment has its place. The design problem is difficult, involving scale, materials, the placement of a new technology and the values inherent in old buildings, including their mechancial systems.

Preservationists and architects show a positive appreciation of energy conservation in old buildings as a measure of a new materialism—a new respect for limited resources. There are compelling reasons to assume that historic preservation and energy-conscious design are complementary. And active solar technology has its place.

Economic Considerations of Solar Systems

James S. Moore, Jr., P.E.

In recent years, the cost of fossil energy for space heating and service water for buildings has grown from a minor expense item to a major burden. For those concerned with old buildings and buildings with historical significance, energy conservation options can be severely restricted because of a variety of circumstances that often occur within such buildings.

For any existing building, there are only two options for reducing energy costs:

1. Substitution of the fuel source currently in use by a cheaper energy source (if available)
2. Reduction in the fuel required to maintain a suitable environment and adequate hot water

Both of these options usually involve some initial cost or capital expenditures to purchase or modify equipment. A designer or building owner thus will seek to arrive at an equitable balance between money invested and the anticipated value of the energy saved by implementing an energy reduction effort. For situations where a new energy system is to be chosen, a similar economic balance must be achieved. Insight into the factors that affect this balance can be achieved by reviewing some qualitative aspects of energy and equipment costs.

The delivered cost of fossil fuels varies widely by region of the United States. For example, the average cost of natural gas used by commercial customers in Delaware is slightly less than \$4.50 per 10^6 Btu. However, in Montana and Nebraska, the cost is less than \$2 per 10^6 Btu. ("Economic Analysis of Commercial Solar Combined Space Heating and Hot Water Systems," Mueller Associates, Inc., September 1980). Similar variations can also be found in the average delivered cost of electricity. In the Northwest, where hydroelectric resources are plentiful, electricity costs a typical commercial customer about 12 mils per kilowatt hour. In New York City, the same type of customer pays about 11½ cents per kilowatt hour ("Commercial Electricity Prices by Cities," *Energy User News*, October 13, 1980).

Opposite: Solar collectors (Mueller Associates, mechanical design) on the west wing of the White House (1792, James Hoban; 1807, Benjamin H. Latrobe), Washington, D.C. (Photo: The White House)

Average Residential Usable Fuel Costs

Natural gas	\$4.61/MBtu
Fuel oil	\$10.90/MBtu
Electricity	\$13.61/MBtu

Source: James S. Moore, Mueller Associates.

However, even with such regional variation conventional energy sources can generally be ranked according to cost as shown in the fuel cost comparison table here.

Solar energy is not shown on the cost line of this table. However, for all energy resources, what we pay for is the equipment to make them usable. While both fossil energy resources and solar energy can be converted to energy usable for space and service water heating, the equipment to convert solar energy to usable heat typically is more expensive than that for fossil energy systems. The operating cost of fossil energy systems, however, is greater than that for solar systems.

Economic Characteristics

Solar systems can be designed to satisfy portions of a building's energy requirement for space heating, cooling and service water heating. The size and complexity of a solar system are determined largely by which of these energy categories the system is to serve and how big a portion it is to meet. System characteristics of a reference solar system for commercial or residential space heating and service water application are shown in the accompanying table. Many variations of solar heating and hot water systems exist.

From an economic perspective, the merit of using a solar system depends on additional factors related to those shown in the table. Geography, energy efficiency of the building, structural quality and configuration, aesthetics, historical value and architectural character are among the most important for old buildings.

Under conditions like those described in the table, the economic return on investment for a solar energy system is influenced by weather patterns, the backup energy system used and local energy costs. Many of the important economic considerations shown in the table represent 1980 conditions and are changing rapidly.

The energy that can be collected to displace the consumption of conventional energy sources varies geographically, with patterns determined by weather factors. Given these weather patterns and knowing local energy costs, the return on investment for commercial solar energy systems for space heating and hot water can be calculated.

The sequential priority of various en-

Solar Energy Systems: Economic Characteristics

System size and solar fraction	
Residential hot water	60 sq. ft./varies
Residential heating and hot water	varies/50%
Commercial	1,000 sq. ft./50%
System installed cost	
Residential hot water	$50/sq. ft.
Residential heating and hot water	$41/sq. ft.
Commercial hot water	$43/sq. ft.
Commercial heating and hot water	$50/sq. ft.
System life	20 years
Down-payment fraction	0.10
Loan interest rate	0.10
Loan life	
Residential heating and hot water	20 years
All other system types	10 years
General inflation rate	0.08
Fuel escalation rate	0.07
Operating energy fraction	0.06
Annual insurance and maintenance cost fraction	0.015
Annual property tax fraction	0.0
Federal tax credit	
Residential hot water	40%
Residential heating and hot water	up to 40%
Commercial hot water	25%
Commercial heating and hot water	15%
Depreciation period (commercial systems)	10 years
Depreciation method	sum-of-years-digits and double declining balance
Income tax rate	
Residential	30%
Commercial	50%

Source: "Summary Report: Economic Analysis of Residential and Commercial Solar Heating and Hot Water Systems," by Mueller Associates, Baltimore, Md., September 1980.

ergy conservation options can be important to consideration of an investment in solar energy. Most architectural and many mechanical and electrical modifications are less expensive than a solar energy system and offer better returns on investment. It is thus usually advisable that such modifications be implemented before investing in a solar energy system. For example, consider a system with the potential to save 200,000 Btu/ft.^{2} per year, which relates to a 50-percent solar fraction (i.e., the solar system is designed to provide 50 percent of the building's space heating and service water energy needs). If this same system were to be considered for application to a similar, although uninsulated building, the solar fraction could be reduced to around 25 percent. While the system would show a similar return on investment in either case, the relative effect on the owner's operating budget is decreased significantly. A reduced interest in making the investment in solar energy thus can be anticipated. With old and historic buildings, insulation of walls and roofs, addition of glazing layers, weatherstripping and improvements to mechanical and electrical systems usually should be considered before an active solar energy system.

Structural and Historical Factors

The return on investment for the example can be altered significantly for specific applications by factors such as building orientation and structural quali-

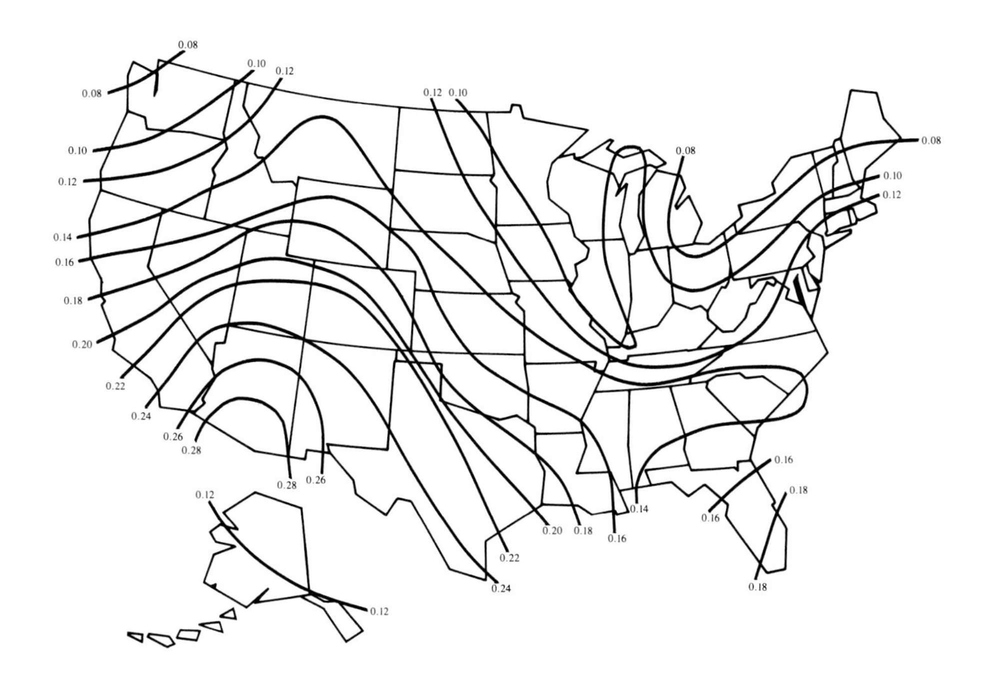

Regional variations in energy savings for residential solar heating and hot-water systems in million Btu/ft^2, for new construction. (Drawing: Mueller Associates)

Return on investment for residential solar heating and hot-water systems with electric resistance auxiliary, for new construction. (Drawing: Mueller Associates)

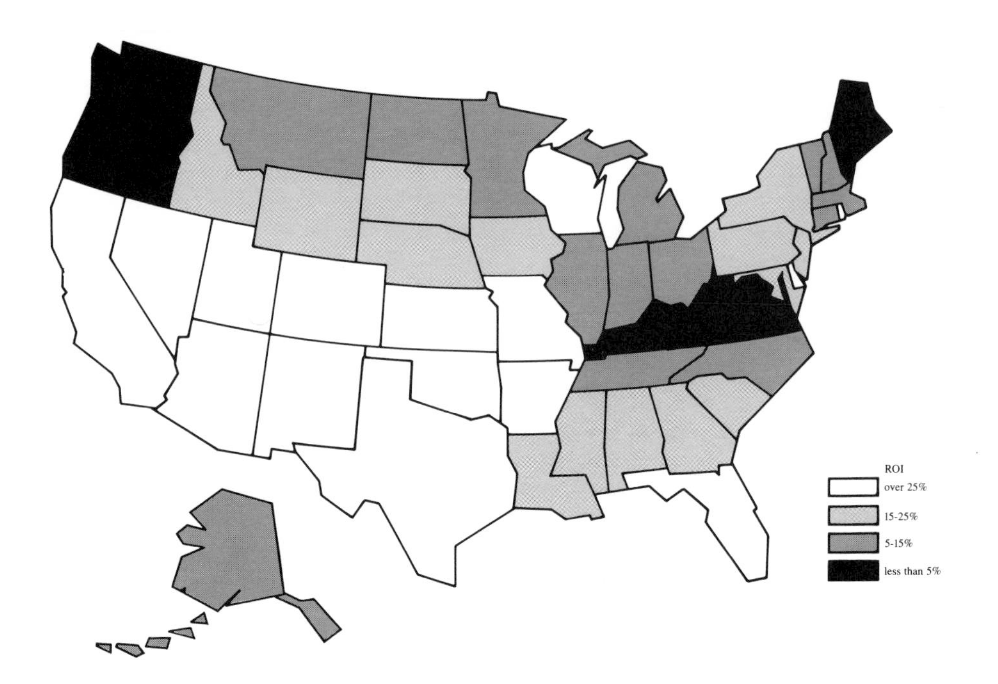

ty. These factors can affect the design of a solar system's collector array. While roof mounting is the most commonly preferred approach, other options may be required for existing buildings. The roof area and configuration (flat versus pitched) can affect placement of the system's collectors. Additionally, the added weight of the collectors and new support structure can exceed the bearing limits of the building's roof support system. For old and historic buildings any one of these problems could prohibit roof mounting of the collectors. For such situations, installation on nearby land is another option. However, any of these configuration or structural quality considerations can measurably increase the system cost and hence decrease the return on investment.

For old and historic buildings historical and architectural characteristics also increase the complexity of integrating a solar energy system. For specific situations such factors can cause anything from a minor impact to a legal prohibition against a solar retrofit. This range of effects can be illustrated in several ways.

On a building with a south-facing roof on the back side, it may be possible to install collectors without having a major architectural effect on the building. At the other extreme are situations in which solar collectors cannot be installed without producing a major visual impact (e.g., a south-facing mansard roof on the front of the building) or in which architectural alterations are forbidden. For buildings in which the interior spaces are as historically important as the exterior, wall insulation and extra glazing may be unacceptable. Solar energy utilization then could be an option for incrementally reducing the building's fossil energy consumption.

The service water heating system for the White House is an example of such a solar application. For this building, major architectural modifications are unacceptable because of its historical importance. However, a solar energy system offered an opportunity to reduce fossil fuel consumption. A feasibility study of a wide variety of applications was conducted. Preliminary layouts, cost and performance estimates for various heating, cooling and service water heating loads were completed. Remote as well as roof locations for the solar collector array were considered. A 600-square-foot collector array deployed on the roof of the Cabinet room in the west wing was selected. This location has minimal architectural impact. The system is integrated with the conventional water heater. Through the use of solar collector performance specifications that allowed selection based on the lowest cost per unit of energy delivered, economical performance was considered. The White House solar energy system strikes a balance among considerations of economics, energy conservation and respect for the character of an old and historic building.

Tax incentives also exist to promote the use of solar energy in commercial and residential buildings. Programs and incentivies available to encourage investment in historic buildings also could be applied to solar energy investments.

In general, solar energy systems can be economically attractive investments, especially when competing against electric resistance heating. However, site-specific factors can significantly alter both capital costs and the energy effectiveness of a system. For old and historic structures architectural and structural characteristics and legal design restrictions can have a range of effects from changing capital costs to barring the integration of a solar system. Alternatively, for buildings where both the interior and exterior are historically significant, the use of solar energy can be one of the few options available to reduce fossil energy consumption.

Solar Alterations to Historic Buildings: The Legal Issues

Frank B. Gilbert

"Now no one, not even a historic district commission, can be opposed to solar heat these days. You have to be for it. Any other position is downright unpatriotic."

That comment from Robert Crouter, a member of the historic district commission in Amherst, N.H., illustrates a new problem for persons administering landmark and historic district laws. In his town across from the common, there is a handsome, 1806 Federal-style building that has been a bank and the parsonage for the Congregational Church. During fall 1979, the present owner put in his front yard a solar collector that Crouter refers to as "a giant trampoline-like contraption of reflective black plastic. The kindest description I have heard of it is that it just looks awful."

The historic district commission had not approved the installation of the collector, and people in the town now expected the commission to handle the problem. After being used for one winter, the solar collector was removed and has so far not reappeared.

The existence of the historic district contributed to this result, because the designation of the area gave validity and strength to the objections in the town. No formal steps were taken by the commission during the controversy. Because of this experience, the commission is now looking for guidelines on the use of solar energy within the historic district.

As energy costs continue to rise, more and more owners will want to use solar energy equipment on their property. This machinery may clash with the historic buildings.

Historic preservation became a factor in American life because of the activities of persistent and idealistic individuals and groups, and now other determined and idealistic persons are promoting the value of solar energy. For preservationists, the current situation presents important opportunities.

For one, historic buildings are regularly adapted to be used for current needs. That philosophy appears in much of the writing about historic preservation. Under that principle, appropriate ways ought to be found to satisfy the requirements of owners for solar energy.

For another, historic preservation laws have gained acceptance as a reasonable regulation of property for a permissible purpose. Preservation legislation has been enacted in more than 600 cities and towns. Under many of these laws, municipal commissions issue certificates of appropriateness permitting owners to make changes to designated landmarks or buildings in historic districts.

The constitutionality of historic preservation was declared by the U.S. Supreme Court in the Grand Central case in 1978, which included language supporting government regulation for aesthetic purposes:

> [T]his Court has recognized, in a number of settings, that States and cities may enact land use restrictions or controls to enhance the quality of life by preserving the character and desirable aesthetic features of a city. . . . [*Penn Central Transportation Co.* v. *City of New York,* 438 U.S. 104, 129 (1978)]

Design review has had a beneficial effect in many old neighborhoods. Now

Opposite: 1837 house in the Mystic Historic District, Groton, Conn., whose owners have sued to retain solar collectors that were installed on the front roof without approval. (Photo: Connecticut Historical Commission)

Amherst, N.H., street bordering the town common. The Federal-style house (1806) at left has a solar collector in the front yard (since removed).

Detail of the solar collector, described as a "giant trampoline-like contraption." (Photos: Robert Crouter)

preservationists will have to educate the solar energy advocates about the procedures leading to approval of their alterations. It is worth noting that previous preservation controversies have often involved demolition threats or alterations that were easy to condemn.

As responsible government agencies, landmark and historic district commissions start with a willingness to approve thoughtful applications that are submitted to them. A New York state court pointed out:

> The regulations do not require that the development of the area remain static but only that the development be consistent with the established nature of the district. It is important that any proposed change should strengthen, not weaken, the character of the district, of course. . . . [*Zartman* v. *Reisem,* 399 N.Y.S. 2d 506, 510, App. Div., 4th Dept. (1977)]

The opinion upheld the decision of a local preservation commission to permit the construction of a tennis court in a backyard within a historic district.

In the administration of historic preservation laws, one difficulty involves inadequate applications from owners—submissions that do not give enough details and do not go into the setting of the proposed changes. In addition, sometimes work is completed, and commissions are then asked to give their approval to dubious changes to buildings. As a practical matter, it is hard to undo work that has been completed at considerable expense.

Both situations point up the importance of preservationists' actively explaining to all owners the review requirements contained in these local laws. The experience under landmark and historic district laws indicates that commissions, when they act in a timely manner, are able to modify the design of alterations submitted to them.

In major controversies, commissions

have also been successful when they reject the plans of owners; there have been some dramatic victories for preservation when demolition applications were turned down. Nevertheless, the proposed alterations of owners are usually accepted in some form by the commissions.

Influencing Design Changes

It is important to understand these principles in relation to solar energy proposals. Preservationists will usually be able to influence the use of solar energy in landmarks and within historic districts. In a few cases there may be valid reasons to turn down a specific application, but local preservation commissions will not be able to ban solar energy.

At times residents react strongly to proposed changes in their neighborhood. Landmark and historic district commissions should expect some opposition to solar energy devices.

Commissions may obtain some guidance from three recent court decisions involving solar energy, although the cases did not involve historic buildings. In addition, preservationists have had some instructive experiences relating to solar energy, but there are no court decisions in these matters yet.

In Mamaroneck, a suburb of New York City, a court decision gave owners permission to construct a solar hot-water system for which panels were to be placed on the roof of their house. [*Katz* v. *Bodkin,* No. 3312/79, Supreme Court, Westchester County, N.Y. (1979)] Initially, the application had been denied by the local building department because the zoning ordinance limited mechanical equipment to 10 percent of the roof and the solar collectors would cover 20 percent of this section of the roof. The owners, who had a $400 grant from the U.S. Department of Housing and Urban Development for installation costs, then appealed to the zoning board of appeals for a modification of the provision containing the percentage limitation.

Silhouettes of the solar collectors would be visible from a distance. The owners' contractor "testified that due to tree shading, only the higher part of petitioners' roof was adequate for the proper functioning of the solar panels and no contrary evidence on this point was adduced."

The zoning board of appeals turned down the application, giving the following reasons (which were later held by a court to be insufficient):

1. Applicant failed to show that the solar collectors could not be located elsewhere on his dwelling where the same efficiency would be provided and be more aesthetically pleasing to the neighborhood.
2. Applicant failed to show that energy could be conserved by other construction methods.
3. Petitioners' land presented no special circumstances or conditions peculiar thereto.
4. Facts and circumstances claimed by petitioner to entitle him to the variance are not such as would deprive him of the reasonable use of his land.
5. The granting of the variance would not be in harmony with the general purposes and intent of the zoning ordinance and not aesthetically in keeping with the residential neighborhood.

The owners modified their application and also submitted consents from 26 of their neighbors, but the appeals board did not reopen the case. At this point the dispute was taken to court.

In finding for the property owners, the court interpreted the zoning ordinance to call for the percentage to be calculated on the basis of the entire roof rather than the section with the solar panels. The court went on to say:

> [I]n this day of what for better expression may be termed the energy crunch, the purposes of restrictive zoning must, to some extent, give way to declared policy of governments to conserve energy in all ways possible yet consistent with environmental standards.
>
> Zoning regulations must be enacted by those to whom such responsibility is assigned in a manner consistent with the promotion of 'health and general welfare' and with 'reasonable consideration, among other things, as to the character of the district. . . . and with a view to conserving the value of buildings and encouraging the most appropriate use of land throughout such municipality' (Town Law, Section 263).
>
> In the accomplishment of the above, it is incumbent upon the zoning agency to adopt an attitude other than an ostrich head-in-the-sand approach, especially when adoption to changing scientific advances follows and complies with national and state interests in energy conservation. It has been said that 'our increasing dependence on foreign energy supplies presents a serious threat to the national security of the United States and to the health, safety and welfare of its citizens' and that, further, 'the mass production and use of equipment utilizing solar energy will . . . promote the national defense.'

The opinion was quoting from two laws passed by Congress, the Department of Energy Organization Act and the Solar Energy Research, Development and Demonstration Act. In addition, the court referred to the New York State Energy Law, which, in section 3-105(2), "imposes upon municipalities the duty to review their rules and regulations and to not only make them consistent with the State declared policy for energy conservation but, where inconsistent, to make necessary changes to comply with the Act's stated purposes." Other states, such as California, have passed laws encouraging solar energy.

Court Permission

In Valencia, Calif., a court overruled the decision of a homeowners association and permitted the installation of solar collectors that were visible. [*Kraye* v. *Old Orchard Association*, No. C 209 453, Cal. Superior Court, Los Angeles County (1979)] These collector plates for a solar water heater would be seen on the roof; thus, they violated the covenants for the property that were recorded more than 10 years before the owner's application to the architectural committee of the association.

As the dispute was being decided by the court, a state law was passed invalidating most covenants restricting the installation of solar energy systems (Cal. Civ. Code Section 714). The law does permit "reasonable restrictions on solar energy systems," which are defined as "those restrictions which do not significantly increase the cost of the system or significantly decrease its efficiency, or which allow for an alternative system of comparable cost and efficiency."

In its ruling, the Los Angeles court went beyond the new state law and declared that the rejection of the roof-top installation was in violation of the public policy of the state.

A different result was reached by the court in the third case, in which the location of a solar collector was turned down. [*D'Aurio* v. *Board of Zoning Appeals*, 401 N.Y.S. 2d 425 (1978)] The decision illustrates that an owner may ask for too much. In Colonie, a town near Albany, N.Y., the zoning required front yards 50 feet deep and free of structures.

The court initially made a concession favorable to the owner: "Although there

Lawton House, East Haddam, Conn., with a representation of a proposed solar collector resembling a cellar door off the side porch. (Photo: Wilson H. Brownell)

appears to be sufficient space at the rear of the lot for the installation of a solar heating unit in compliance with the zoning ordinance, the unit apparently would not be most effective there." The judge went on to describe the proposed unit: "It is placed on skids and the solar heating panels are 8 feet in height. The unit is 8 feet wide at the base and is 20 feet 9 inches in length." In reaching its decision, the court used a standard that would also be applied in judging the actions of landmark and historic district commissions: "The court concludes that, with respect to petitioner's application for an area variance, the respondent did not act in a manner that was in any way arbitrary, unreasonable, irrational or indicative of bad faith."

Review Board Experience

Historic district commissions are getting some experience with solar energy issues. In Charleston, S.C., where America's first local historic district was established in 1931, the board of architectural review has approved solar hot-water systems within its district. Some minor adjustments were made in proposals submitted; for example, on a flat roof, collectors were moved so that they would not be visible from the street, and in another situation the backs of the collectors and the piping were painted to match the color of the roof.

In Connecticut, the East Haddam Historic District Commission in December 1977 turned down the installation of solar heating panels on two houses owned by the Goodspeed Opera House. The vote was 3 to 2 on each building, following three meetings about the plans. The original proposals called for panels on the roofs of both houses, but one of the installations was changed to resemble a small greenhouse placed next to the building like a cellar door with shrubs planted nearby.

The rejection was controversial in the

Boardman House, East Haddam, Conn. A rendering shows the proposed placement of the collectors on the roof at right. (Photo: Wilson H. Brownell)

community, and the owner initiated a lawsuit. As the case developed, it appeared that the presentation by the owner to the commission could have been more thorough; photographs of the houses with an artist's rendering of the two installations were circulated after the commission's decision. The commission itself did not have a transcript of its meetings. Depositions were taken of commission members on their reasons for rejecting the proposal.

The lawsuit eventually was withdrawn. More than three years after the first decision, the owner is planning to present his plans again to the commission.

Elsewhere in Connecticut, a lawsuit has been brought by a family trying to retain solar panels that were installed without permission on its 1837 Greek Revival house. [*Gunther* v. *Historic District Commission*, No. 56891, New London Superior Court, Conn. (1980)] The owner, William Gunther, blamed his contractor for the failure to get a certificate.

After the application was made, the Groton commission approved the use of solar panels in less prominent locations on a wing of the house or on the grounds. The commission made these stipulations:

a) The panels will be installed either on the westernmost ell of the house or on a ground mounting contiguous to either ell and as close to ground level as possible.
b) If the panels are installed on the ell, they shall be mounted directly on the roof and at the same pitch as the existing roof line.
c) The roof of the main house, on which the previously unauthorized panels were installed, shall be restored to its original condition and appearance.

Gunther says that the conditions imposed by the commission will keep him from using solar energy in his house.

In its ruling the commission stated that solar panel installations were desirable where such installations can be accomplished with:

1. Minimum visual impact on the surrounding neighborhood.
2. No change to the appearance of the historic structure other than from the panels.
3. Installations which do not permanently change the significant architectural features of the building.
4. Location at ground level, or failing to secure adequate solar benefits at this location, then installation on less visible and lower elevation ells or wings.
5. Positioning so as to match the existing slope of the roof as well as possible.
6. For ground mountings, use of landscaping and appropriate fencing to minimize visual impact and promote safety for persons on the property.
7. Use of panel structures and textures which blend with existing building structure.

The owner argues that federal and state laws have established a priority for reasonable use of solar energy. There also are factual issues in the case: whether the alternative locations will receive enough sun, whether the ell roofs are strong enough to support panels, and whether a ground location would be dangerous for other persons and thus an attractive nuisance and an uninsurable risk. A decision in the case is pending.

Current litigation in West Barnstable, Mass., is involving a commission in technical questions about solar energy. (Courts reviewing commission decisions regularly receive testimony on economic questions; for example, can a property threatened with demolition be put to a reasonable economic use?) While a commission's decision will be based on architectural and historical reasons, applicants may have put into the record engineering information that will be helpful to their position, as it was in the Mamaroneck litigation described previously. A commission may need expert help to evaluate solar testimony and to place a full picture into the record.

In the Massachusetts case, the historic district commission turned down an application because "the solar panels when placed on the roof would contrast sharply with the cedar shingle roofs that presently exist on all buildings within the sub-division." In preparing for the trial, the commission's lawyer asked whether six panels on the roof would actually provide heat as well as hot water or whether three panels would be sufficient for hot water. The cost of different design solutions also was an issue in the litigation. [*Roehlk* v. *Old King's Highway Regional Historic District Commission*, No. 15887, Barnstable District Court, Mass. (1980)]

Guidelines are an important tool for commissions in educating owners and in reaching decisions on applications. A February 1980 report of the Central Naugatuck Valley Regional Planning Agency in Connecticut included these recommendations:

> In determining the appropriateness of solar energy systems on historic buildings, historic district commissions should consider (1) the visibility of the system from the public way, (2) the extent to which the solar energy system maintains the historic building as a continuously useful and usable structure, and (3) the harmony of the system with the design of the building.

The report also refers to the opinion of

Interior of Maxamillian's Restaurant in the Old Town Historic Zone, Albuquerque, N.M. Originally a two-story adobe and brick exterior courtyard, it was enclosed with four glazed sawtooth clerestories that provide heating and cooling. The design (Mazria and Associates) was approved by the review board. (Photo: Tom Gettings)

the Woodbury Historic District Commission:

> Collectors located in backyards, backroofs, sideyards (screened from public view), or roofs slanting to the side (rather than toward the road) would not be considered as posing serious threats to the integrity of the historic district. . . . The Commission members agreed that any changes made on the interior of the house, such as water walls, or other passive design modifications, would not require Commission approval.

Commissions should be aware of the need to give reasons for their decisions. Recently, a court required a commission to reconsider its decision denying permission to install redwood siding on a brick house within a historic district. In its opinion, the court declared:

> The record must disclose the facts on which the Commission acted and a statement of the reasons for its action. Without such a record the reviewing court cannot perform its duty of determining whether the action of the Commission was arbitrary or capricious. [*Fout* v. *Frederick Historic District Commission*, No. 4005, Frederick County Circuit Court, Md. (1980)]

Commissions do have the responsibility of reviewing changes to all property within a historic district even when the site does not contain a historic building. In 1979 the North Carolina supreme court approved the designation of a historic district in Raleigh and the inclusion of a vacant lot within the district:

> It is widely recognized that preservation of the historic aspects of a district requires more than simply the preservation of those buildings of historical and architectural signifi-

Adobe house, Albuquerque, N.M., showing glazed Trombe wall enclosing the once-open portal. The house has been accepted on the state landmarks register, and the state landmarks review committee did not object to the solar wall. (Photo: Christopher Wilson, Historic Landmarks Survey of Albuquerque)

> cance within the district. . . . This 'tout ensemble' doctrine, as it is now often termed, is an integral and reasonable part of effective historic district preservation. [*A-S-P Associates* v. *City of Raleigh*, 258 S.E. 2d 444, 451 (1979)]

As landmark and historic district commissions decide applications submitted to them, they may be helped by the following principles:

1. Owners have needs that should be met. This approach has to be the starting point in order to make historic preservation work. Under the provisions of historic preservation laws, owners are asked to take a number of steps over a limited period of time so that their needs and

Harwood House, Spring Garden Historic District, Philadelphia. Rebuilt from its historic shell, the house was designed (1979, Stephan J. White) for energy efficiency with superinsulation and solar features, which are not visible from street level.

plans will be understood.

2. People want immediate approvals of their plans. There is an emotional point related to this principle: A person's desire for a permit to do some work increases dramatically when the work has already been done without the required approvals. In the field of historic preservation, owners have the opportunity to start the approval process whenever they are ready. Owners bring in their plans, and the commissions then review them. In fact, landmark and historic district commissions would like to talk informally with owners before architectural drawings are made, materials are purchased and contracts are signed.

3. Historic preservation is a search for alternatives. There is a better way to make changes in buildings, and this can be discovered by developing the facts about an owner's plans.

4. An architectural solution can be found for most controversies about historic buildings. Up until a few years ago, people did not look closely at old build-

The solar greenhouse and hot-water collector panels on the roof of the Harwood House can be seen from a roof-top across the street. (Photos: Stephan J. White)

ings and their potential. Buildings were demolished before there was a new use for the site. With historic properties that remained standing, contractors were free to do many strange things to old buildings. Bizarre alterations were made to the first floors of old buildings, both commercial and residential structures. Now a vacuum has been filled because of the preservation laws that have been passed and the procedures under those laws. At monthly meetings of landmark and historic district commissions, owners discussed the design, the materials and other elements of their alterations, and members of these boards also ask about the buildings to the left and right of the property under consideration and those across the street. It is useful to see changes in the context of a neighborhood.

5. When the choices are developed properly, landmark and historic district commissions will reach responsible decisions that will be fair to owners.

4. Afterword

An Opportunity for Preservation Leadership

Paul Goldberger

Over the last few years, preservation has come more and more to resemble motherhood (or at least what motherhood used to mean to Americans). Energy conservation is achieving the same sort of mythical status. And joining together preservation and energy conservation is something of a natural.

But in the sense of the opportunity it offers, the energy situation is better for preservation than that of another crisis, the 1974-75 recession, which sharply cut back construction in much of the United States and thus was indirectly a great aid to preservation. Then, the benefits that accrued to the preservation movement from the slowdown in construction seemed almost sinister, as if the profit came from the rest of the country's problems.

Preservation: A Real Solution

Energy is a bit different. While it is, indeed, a grave problem, there is no sense that the preservation movement is at odds with the rest of the country over energy. It is more a question of the preservation movement's being able to offer a solution, something it was unable to do for the recession, when it merely could see its own impulses, shall we say, strengthened. But preservation does offer at least a partial solution to the energy crisis. The preservation movement thus has been wise to exploit and make known the connections between preservation and the national interest in energy.

One important issue that has been raised is the fact that the benefits of preservation and energy conservation frequently are societal and not personal. That is a point well taken because, for example, there is no personal gain at all for an office renter who is paying on a lease and saves energy. Until a system in which energy benefits are forced into the economic system for the general benefit, much as the Tax Reform Act of 1976 has done for rehabilitation projects, that situation will remain largely unimproved.

Many statistics bearing on energy conservation benefits have been quoted, and it is appropriate to review briefly a few of the more staggering ones. Of all the energy in the United States, 38 percent is consumed by buildings. Of that energy, 41 percent probably is wasted or unnecessarily expended. By 1990, it may be possible to save 8 quadrillion Btu's per year through adequate retrofitting. And then there is that extraordinary fact of 450,000 Btu's somehow hidden in a gallon of paint, a fact my hardware store must have been aware of at the latest price rise! These statistics continue to shock, in spite of the fact that they have been cited again and again.

Richard Stein's work, which is crucial to an understanding of the role of architectural design in energy savings, was ignored at the beginning, began to get a little attention a couple of years ago and only now seems to have achieved a status of common wisdom.

It must be remembered that the embodied energy in each existing building is like a bank deposit. It is not available to be withdrawn, but the interest it produces is there to be used. If the building is demolished, the deposit is, in effect, forfeited, and the entire capital represented by that deposit must be replaced merely to get back to the same level of interest or income—plus, of course, there is the added cost of demolition energy.

Opposite: Williamson Hall (1977, Myers and Bennett/BRW), University of Minnesota, Minneapolis, is energy conserving while protecting the view of the old library. (Photo: Myers and Bennett/BRW)

Elevation and roof view of proposed energy-efficient infill construction for international student housing adjacent to the University of Minnesota, Minneapolis. The proposal was a winner in the 1980 National Student Design Competition, "Design + Energy" and featured reuse of existing houses on the site. (Drawings: James Rasche, University of Minnesota)

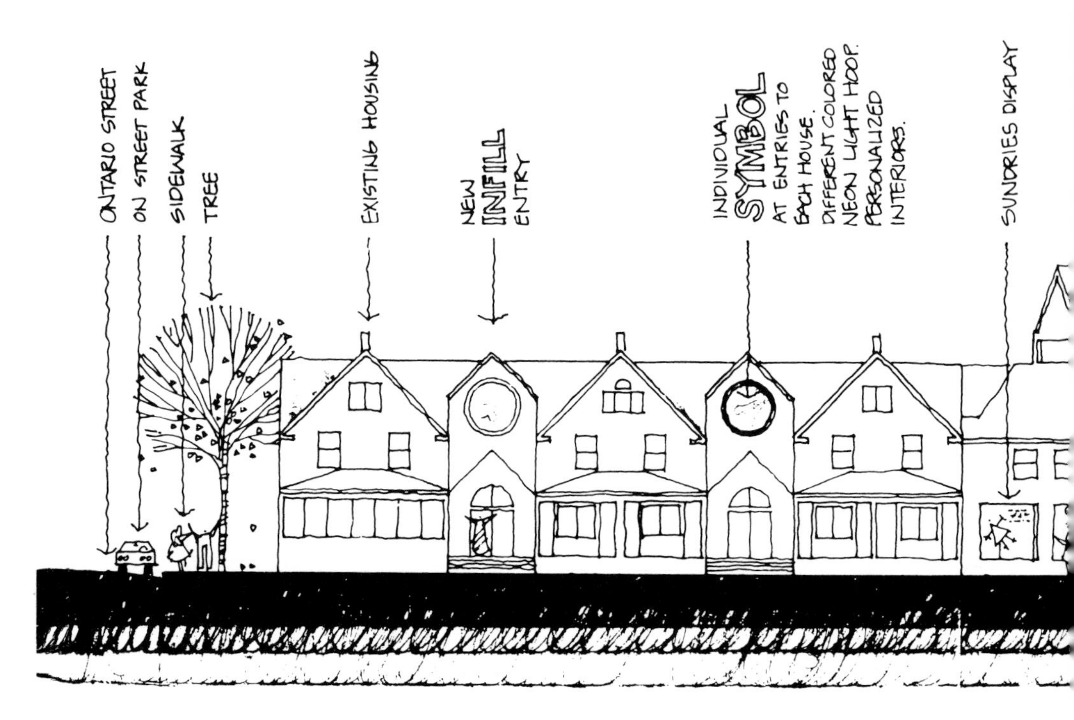

Reaction against Unthinking Changes

When one thinks of it that way, one begins to wonder why there is new construction at all, why anyone has ever built a new building. In that light, it is important not to forget the realities and not to fall into the trap of sentimentalizing historic preservation. Often, old buildings do not serve current needs. Communities must live and grow, and this must bring change. The problem is that all of this has happened too much and too fast. The preservation impulse is in large part a reaction against this excessive, rapid and often vicious pace.

It is also a reaction against the nature of the changes to the environment that have come in the last couple of generations. As a rule, architecture is not trusted, at least in terms of quality, to replace what is torn down. There is an unspoken belief that things are preserved, not always because we believe they are so good, but because we do not believe that anything that might replace them is likely to be any better or even as good. This is a sad and rather cynical comment on modern culture, but often it is correct, and often, in fact, it is justified.

This is not as much of a digression as it may seem, because this subject ties into larger preservation issues. One prevalent belief, and another that was not spoken explicitly, is the notion that the answer to many problems is conventional buildings better done—old buildings retrofitted or fairly conventional new buildings, the sort of buildings to which we all are accustomed, the sort of buildings with which we grew up, the sort of buildings we all lived in and around and saw every day.

Salvation does not lie in jazzy energy gimmicks or in vast solar collectors any more than it lies in jazzy modern design and splashy glass boxes built all over, indifferent to the layout of the environment. One wonders why the so-called postmodernists in architecture have not seen this connection, where they, too,

"Its recognition of the value and opportunities of incorporating existing buildings into its solution combined an energy awareness with a great sensitivity to the scale and the context of the existing buildings."

Richard G. Stein
Competition juror

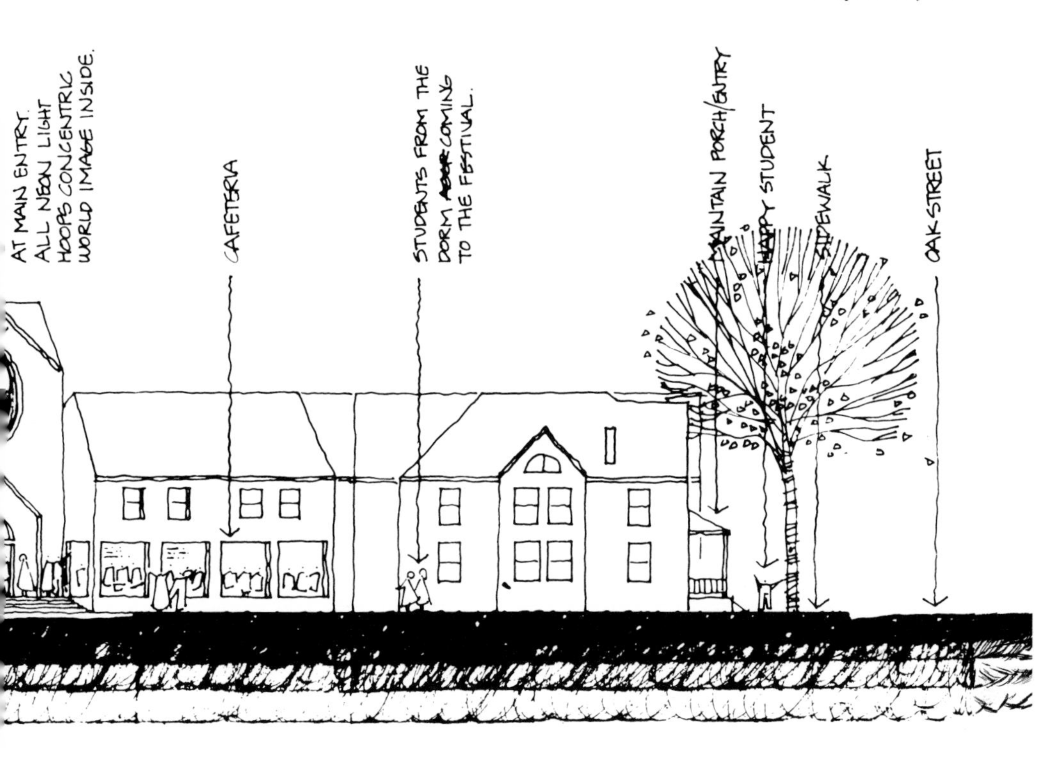

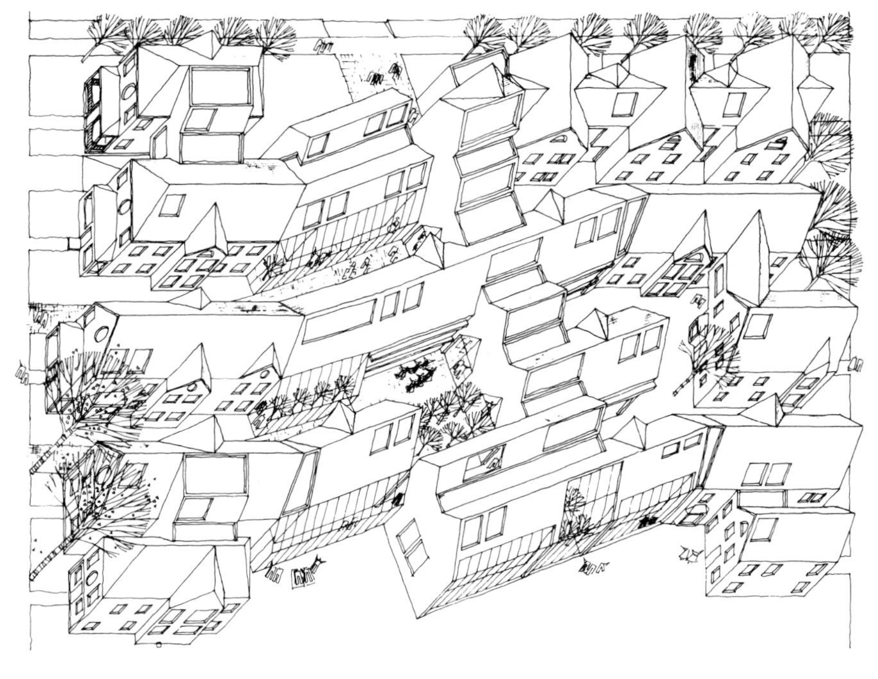

Aerial perspective and gallery section for a proposed Skidmore College visual arts center, Saratoga Springs, N.Y. The winner in the open submissions category of the 1980 National Student Design Competition, the proposal suggested a long, narrow atrium, thermal storage walls and double-glazed skylights. (Drawings: Scott Barton, Rensselaer Polytechnic Institute)

"The resemblance to a New England factory is pleasing, visually attractive and symbolically appropriate to the program."

Paul Goldberger
Competition juror

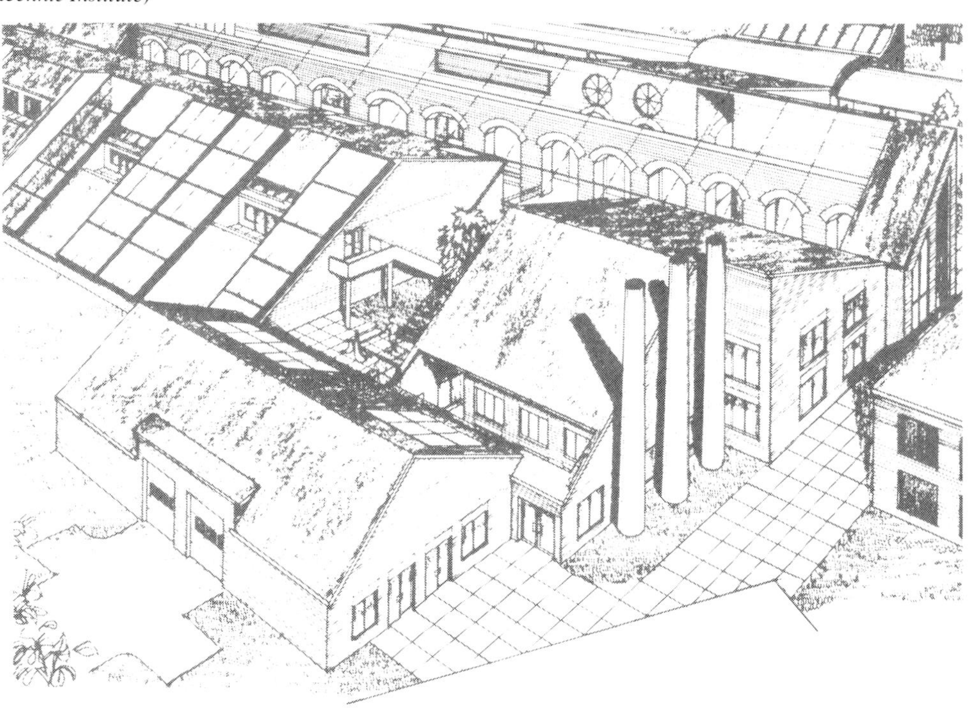

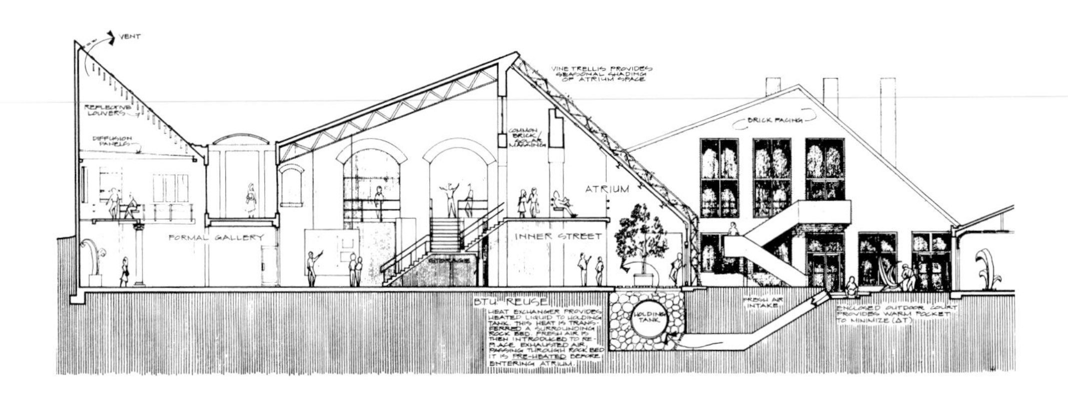

East elevation and ground plan for the Skidmore visual arts center, designed to bring daylight into the painting, gallery and related spaces.

argued against the orthodoxy of modern architecture and the directions in which the standard glass box has moved, and they, too, claim that much is to be learned from the conventional buildings on the Main Streets of every town. Still, they have mysteriously declared energy to be irrelevant to design concerns. It is not so at all.

As Theodore Sande points out earlier in this book, there is a long and honored history of such concern. In fact, one can return to the wonderful quotes he offers from Vitruvius: "We must at the outset take note of the countries and the climates in which [houses] are built. . . . Winter dining rooms and bathrooms should have a southwestern exposure. . . . Bedrooms and libraries ought to have an eastern exposure, because their purposes require the morning light," etc. Put aside the fact that creating for seasonal dining rooms is not the sort of problem that every architect faces today. The importance is in the notions underlying Vitruvius's idea that the integrity or art of design need not be compromised by accommodation to natural factors. If that is good enough for Vitruvius, it ought to be good enough for Philip Johnson.

Reviving an Old Tradition

The point is that there has been a tradition of expectation that these things would be taken into account, rather as Louis Kahn once spoke of the plumbing and the electrical systems and the functional organization of a building, not as something that architects should ignore or consider themselves above, but as elements that should be taken for granted as a crucial part of the design process. The aesthetics, Kahn noted, begin once these are integrated into an overall concept, rather than being seen as an affront or an obstacle to the making of art, as so many contemporary architects, especially in the heyday of the 1960s boom, have tended to consider them.

In his book *Architecture and Energy* (Anchor Press, 1977), Richard Stein wrote

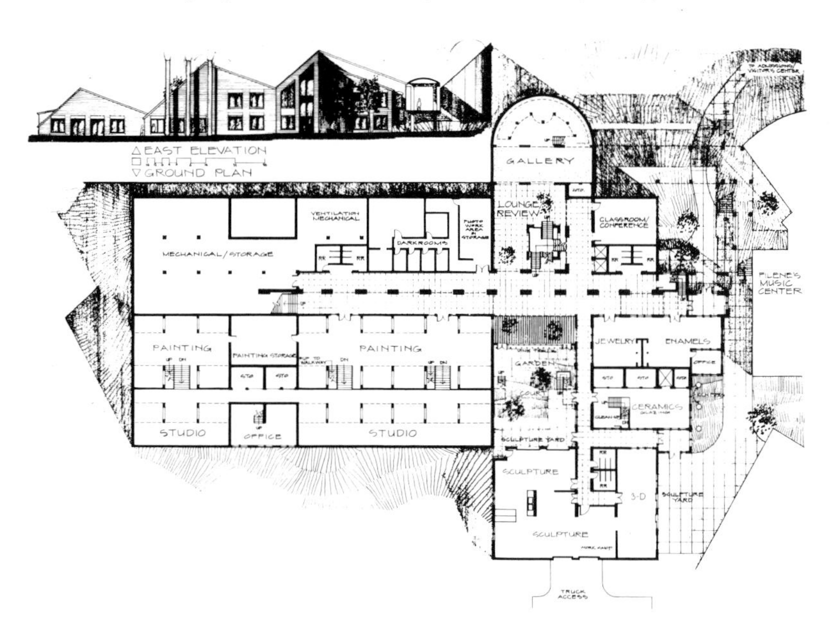

A hypothetical modern Hawaiian solar house using traditional island building techniques. Natural energy-saving features include sun control through deep overhangs, high vented roofs allowing heat to escape and free ventilation through open walls and large windows. (Drawing: Jim Pearson, Historic Hawaii Foundation)

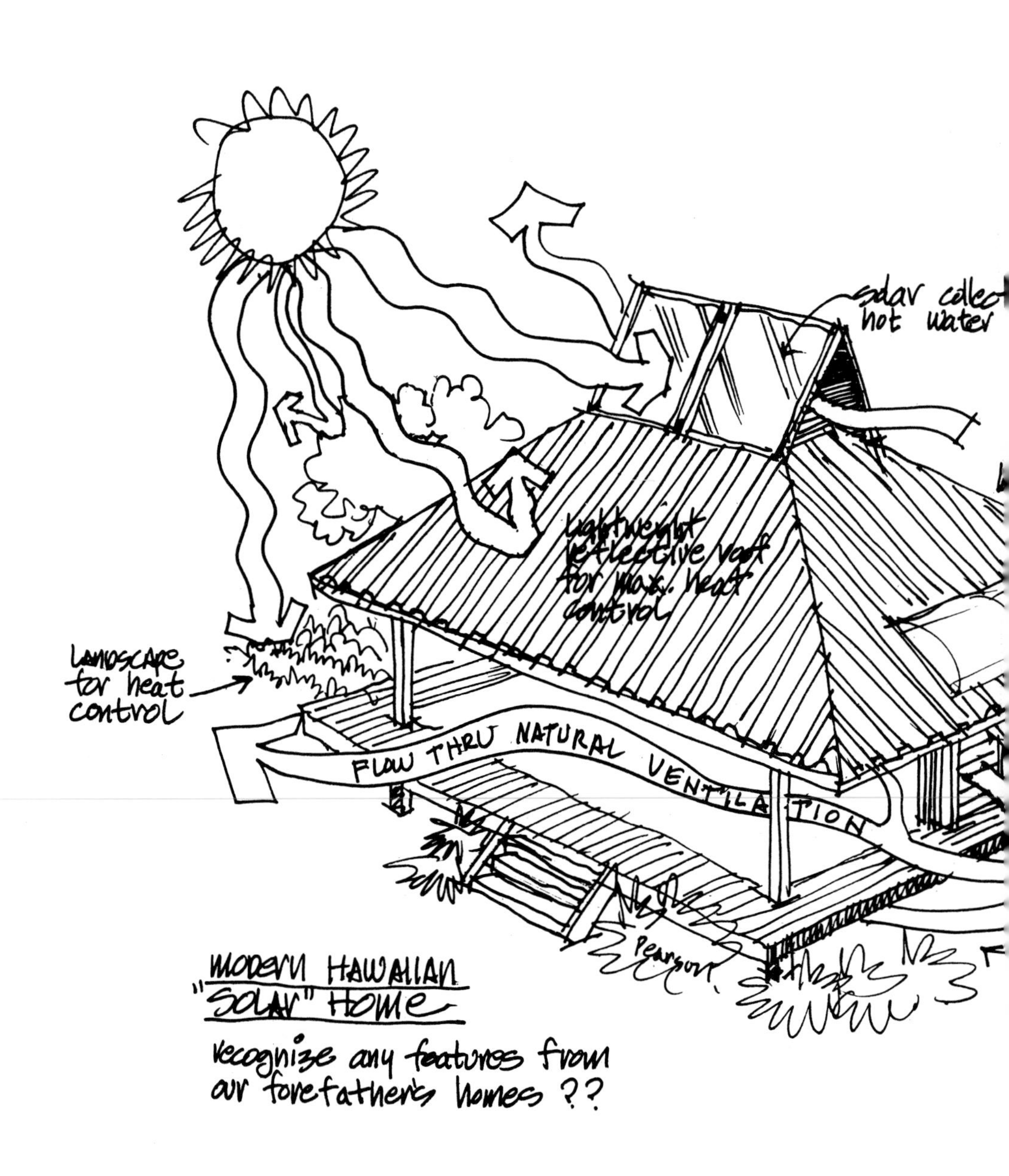

"There are many lessons to be learned from traditional Hawaiian architecture which will allow us to enjoy modern standards of comfort and convenience while cutting way back on energy consumption."

Jim Pearson and Howard C. Wiig
Historic Hawaii News

of the energy problem as being liberating for architects. He envisioned a time when more architectural regionalism would result from accommodation to different climates and local materials—more thoughtful orientation of buildings on sites; more limited and unwasteful use of buildings, both within building plans and in overall complexes; less of the flashy, sexy razzmatazz that often exists for no particular reason; more sensible use of materials, etc.

The culture at large now is more responsive to preservation than ever before. Still, the argument must be made against the preservation reflex, the automatic instinct that everything always must be saved under any circumstances because it is old. Fredric Quivik made a similar point in arguing for flexibility in interpreting design guidelines to prevent a sacred-cow approach when retrofitting buildings. That is not to say that work should be done willy-nilly or indiscreetly or without guidelines; but individual, case-by-case judgments, not formulas, are needed, because formulas never have been the answer, and they will not be the answer now. Of course, most old buildings of landmark quality still are fairly sacred and must be treated carefully, for the preservation impulse, even if it should not be an automatic reflex, is still more right than wrong. Preservation does make sense more often than not.

So energy, like the mid-1970s recession was and perhaps like the new recession again may be, is a great opportunity for historic preservation. But it is a much more positive opportunity than before, the kind of opportunity that can offer leadership and answers to a country, not merely a sense of profiting from misfortune.

Preservationists can show the rest of the country that in the end things are not necessarily so bad, that the seeds of improvement and the way out of a crisis can lie within what we already have.

Glossary

absorber plate Solar collector surface, usually blackened metal, that absorbs solar radiation.

active solar system Uses hardware and mechanical equipment to collect and transport thermal energy for hot water or space heating. Major components are collectors (usually mounted on a building roof) and a separate heat storage unit, from which heat is supplied to the building spaces by a mechanical distribution system.

Btu British thermal unit. Standard measurement of energy content defined as the amount of energy required to heat one pound of water 1°F. Other measurements often are converted into Btu's to express total energy content or use, e.g., one gallon of gasoline is equivalent to 125,000 Btu's. One Btu is approximately equal to the amount of heat provided by burning one kitchen match.

cavity wall Exterior wall, usually masonry, consisting of an air space between an outer and inner section and providing improved thermal insulation.

clerestory Upper window or level of windows designed to admit light into a high-ceilinged room. Clerestories are effective for directing sunlight to strike an interior wall being used for thermal storage.

cogeneration Technique for capturing steam waste rejected when electricity is produced and using it for space heat or residential or industrial hot-water heating. The process saves energy over separate electricity and thermal generation systems. It is most applicable to industrial and large commercial and institutional uses.

collector, solar collector Device for receiving solar radiation. Usually refers to a mechanical component of an active solar system, but may include a passive solar device such as a south-facing window.

conduction Process by which heat energy is transferred through materials (solids, liquids or gases).

conductivity Ease with which heat will flow through a material determined by the material's physical characteristics.

convection Transfer of heat between a moving fluid and a surface, or the transfer of heat within a fluid by movements within the fluid.

design review Process of ascertaining through prior review whether proposed modifications to structures in designated historic districts or other similarly protected areas meet standards of appropriateness.

direct gain Most basic passive solar heating system, in which the space to be heated directly absorbs sunlight and serves as collector, storage space and distribution system in one. The system requires south-facing glazing to collect sunlight and a thermal mass (masonry or water) for absorption and storage.

double envelope Type of prototype new house that is partly a house within a house, designed to allow free circulation of air in a convective loop including a south-facing greenhouse, roof or attic and basement or crawl space.

double glazing A window with two parallel panes of glass or other glazing material with an air space in between designed to increase thermal insulation.

embodied energy Amount of energy, measured in Btu's, represented by the production, delivery and installation of materials in an existing building. The Btu's are often expressed in equivalent gallons of gasoline. The energy equivalent of one gallon of gasoline is required to make, deliver and install eight bricks.

envelope Space occupied by a building as enclosed by its outer walls and roof. Also, the imaginary shape of a building indicating its maximum volume.

flat-plate collector Solar device to absorb sunlight and convert it into heat. Usually consists of a metal panel painted flat black on the side facing the sun, placed in an insulated box and covered with glass or plastic to retard heat loss. The heat is transferred in the collector to circulating liquid or gas, such as air, water, oil or antifreeze, which then transfers or stores the heat for use.

focusing collector Solar device with a reflector that focuses sunlight on a small area for collection. By intensifying the heat, this type of collector allows the storage system to reach higher temperatures.

glazing Glass or plastic covering used to admit light through a window, door, skylight, etc.

greenhouse Freestanding or attached glazed structure that serves as an isolated gain passive solar system. Solar heat is collected through the glazing and stored in a thermal mass such as a mass wall, rock storage bed or water drums. Heat is transferred to the adjacent space through conduction or convection.

indirect gain Passive solar heating system in which sunlight is stored rather than being directly used for space heating. The light strikes a thermal mass such as a thermal storage wall or a roof pond. The absorbed sunlight is converted into heat and transferred into the adjacent space by conduction and convection.

infiltration Flow of air into a room or space through cracks around windows, under doors, in walls, roofs, floors, etc.

insolation Amount of solar radiation striking an exposed surface from direct, diffused and reflected sources.

isolated gain One of three passive solar heating approaches in which solar radiation is captured by a separate space such as a greenhouse or atrium. The system relies on a natural convective loop to transfer heat from the storage area to the space to be heated.

mass wall Masonry wall or water-storage enclosure (metal drums, plastic, concrete) used in an indirect gain solar system to absorb, store and distribute heat. Such walls are usually located at least four inches behind a south-facing glass or transparent plastic enclosure.

microload passive Building with insulation sufficiently airtight (superinsulation) to require only minimal additional passive solar energy and internal heat sources, resulting in a small or micro heat load.

monitor window Opening in a monitor roof, which is a raised section of a roof often straddling a ridge.

passive solar system One in which the building or space is itself the energy collection system and for which no separate mechnical collectors, storage units or distribution elements are used. Thermal energy is distributed by natural means such as radiation, conduction and natural convection. Two basic elements are south-facing glazing for solar collection and thermal mass for absorption, storage and distribution. Three types of system are defined as direct gain, indirect gain and isolated gain.

payback Amount of time necessary to obtain a return through energy savings on the cost of investment in energy conservation measures or devices. A low payback period is 6-9 years; medium, 3-5 years; high, 1-3 years, depending on the cost of the item, the climate zone and temperatures maintained in the building.

phase-change materials Heat-storing substances such as certain salts that are used in solar energy systems. The materials melt and solidify near room temperatures, absorbing a large amount of heat when they melt and releasing it when they become solid.

photovoltaic process Direct conversion of the sun's energy into electricity through the use of solar cells or semiconductors, usually silicon, assembled in a solar array. As sunlight hits the solar cell it creates an electrical current that is transmitted to a battery.

radiation Direct transmission of thermal energy through space.

renewable energy Nondepletable resources such as sunlight, wind, hydropower and nuclear and geothermal energy. Depletable sources of energy include fossil fuels such as oil and coal and natural gas.

retrofit To modify existing buildings to receive contemporary mechanical systems or convert them to new energy conservation systems, including solar techniques, insulation and more economical and efficient heating and cooling systems.

rock storage bed Area of rocks beneath the floor or in a wall of a greenhouse or other passive solar system designed to store heat collected during the day for use at night.

roof pond Water enclosed in plastic bags serving as a thermal mass on a building's roof. They store heat for conduction through the ceiling and radiation to the spaces below.

R-value Thermal resistance, or a measure of the ability of a substance to resist the flow of heat. A higher R-value of an insulation product indicates better insulating ability.

skin Materials used to form a building's exterior wall and to enclose interior space, often used in reference to nonbearing walls.

solar access Exposure of a building or its solar collectors to the sun. A building with limited southern exposure, surrounding trees or other structures blocking receptivity to sunlight lacks optimum solar access. Also, the issue of legal rights to enjoy or maintain such exposure.

solar fraction Percentage of the energy requirements met by a solar collector. A system that captures half the incoming radiation is considered 50 percent efficient, although 55 percent is good under desirable weather conditions.

solar rights Legal issue of assuring access to the sun for light and operation of solar energy systems. Obstruction of access by tall buildings or trees after installation of solar systems diminishes or negates their value. The existence of a common law right to solar access is disputed, but states and localities are beginning to guarantee solar rights through specific legislation. Easements or covenants also are being used to assure solar access.

storm window, storm sash An additional layer of glazing usually placed on the outside of an existing window as protection against cold air.

sun space Solar-heated room or buffer space between a building and the outside such as an attached greenhouse, sun porch or solarium. Sun spaces lower the rate of heat loss from the building because the space becomes warmer than the air outside through thermal absorption.

superinsulation Airtight insulation of a building sufficient to keep it warm with only a minimum of passive solar energy (e.g., through windows) and internal heat sources within the building (from occupants, lights, electrical devices, etc.). Limited auxiliary heat sources such as a heat stove or furnace may be used rarely, if necessary. Also termed microload passive.

thermal mass Materials that store heat for release when needed, such as a concrete floor, masonry wall, rock storage bed and drum or water wall.

Trombe wall Concrete thermal storage wall in an indirect-gain solar energy system. Named for engineer Felix Trombe, who built several houses in France in the 1960s with this type of double-glazed thermal wall approximately two feet thick, painted black to absorb the sunlight passing through the glass. The wall heats the building by radiation and convection.

U-value Measure of the rate of heat loss through a wall, roof or floor. U is the combined thermal conduction value of all the materials in a building section, plus air spaces. The lower the U-value, the lower the heat loss or the higher the insulating value.

vapor barrier Coating, layer of material or space impermeable to moisture penetration that is designed to prevent passage of water or water vapor into a structure or insulation materials.

weatherization All efforts such as caulking, weatherstripping, addition of insulation and installation of storm windows intended to seal a building against cold air, wind, rain and snow.

Selected Bibliography

General

Commoner, Barry. *The Politics of Energy.* New York: Alfred A. Knopf, 1979.

Farallones Institute. *The Integral Urban House: Self-Reliant Living in the City.* San Francisco: Sierra Club Books, 1979.

Fitch, James Marston. *American Building: The Environmental Forces That Shaped It.* 2nd ed. Boston: Houghton Mifflin, 1972.

Hunt, V. Daniel. *The Energy Dictionary.* New York: Van Nostrand Reinhold, 1979.

Olgyay, Victor. *Design with Climate: Bioclimatic Approach to Architectural Regionalism.* Princeton: Princeton University Press, 1963.

Stobaugh, Robert, and Daniel Yergin, eds. *Energy Future: Report of the Energy Project at the Harvard Business School.* New York: Random House, 1979.

Subcommittee on the City, Committee on Banking, Finance and Urban Affairs, U.S. House of Representatives. *Compact Cities: Energy Saving Strategies for the Eighties.* Washington, D.C.: U.S. Government Printing Office, 1980.

Warren, Betty. *The Energy and Environment Checklist: An Annotated Bibliography of Resources.* San Francisco: Friends of the Earth, 1980.

Werth, Joel T., ed. *Energy in the Cities Symposium.* American Planning Association Planning Advisory Service Report No. 349. Sponsored by U.S. Department of Housing and Urban Development. Chicago: American Planning Association, 1980.

Energy Conservation

American Society of Heating, Refrigerating and Air Conditioning Engineers. *ASHRAE Handbook of Fundamentals.* New York: Author, 1977.

Blandy, Thomas, and Denis Lamoureux. *All Through the House: A Guide to Home Weatherization.* New York: McGraw-Hill, 1980.

Booz, Allen and Hamilton. *Assessing the Energy Conservation Benefits of Historic Preservation: Methods and Examples.* Advisory Council on Historic Preservation. Washington, D.C.: U.S. Government Printing Office, 1979.

Burch, Douglas M. *Retrofitting an Existing Wood-Frame Residence for Energy Conservation: An Experimental Study.* National Bureau of Standards, U.S. Department of Commerce. Washington, D.C.: U.S. Government Printing Office, 1978.

Dubin, Fred, and Gary Long. *Energy Conservation Standards. New York: McGraw-Hill, 1978.*

Environmental Study Conference, U.S. Congress. *Guide to Federal Energy Conservation Assistance.* Grace Malakoff, ed. Washington, D.C.: Author, 1980. Available from U.S. senators and representatives.

Langdon, William K. *Movable Insulation.* Emmaus, Pa.: Rodale Press, 1980.

Massachusetts Energy Office, Xenergy, Inc., and Center for Information Sharing. *Reducing Energy Costs in Religious Buildings.* Boston: Center for Information Sharing, 1978.

Moffat, Anne Simon, and Marc Schiler. *Landscape Design that Saves Energy.* New York: Quill Books, Morrow, 1981.

Naumann, Herbert. *Natural Cooling for Homes: Low Energy Concepts.* Butte, Mont.: National Center for Appropriate Technology, 1979.

Neblett, Nathaniel P. *Energy Conservation in Historic Homes.* Washington, D.C.: Historic House Association of America, 1980.

Nielsen, Sally E., ed. *Insulating the Old House.* Portland, Maine: Greater Portland Landmarks, 1977.

Old-House Journal. Special energy issue. September 1980.

Peterson, Stephen R. *Retrofitting Existing Housing for Energy Conservation: An Economic Analysis.* Building Science Series 64. National Bureau of Standards. Washington, D.C.: U.S. Government Printing Office, 1974.

Rothchild, John, and Frank F. Tenney, Jr. *The Home Energy Guide: How to Cut Your Utility Bills.* New York: Ballantine Books, 1978.

Rubin, Arthur I., Belinda L. Collins and Robert L. Tibbott. *Window Blinds as a Potential Energy Saver: A Case Study.* National Bureau of Standards. Washington, D.C.: Author, 1978.

Shurcliff, William. *Thermal Shutters and Shades.* Andover, Mass.: Brick House Publishing, 1980.

Sizemore, Michael, Henry Clark and William Ostrander. *Energy Planning for Buildings.* Washington, D.C.: American Institute of Architects, 1979.

Smith, Baird M. "Conserving Energy in Historic Buildings." Preservation Brief No. 3. Technical Preservation Services Division, U.S. Department of the Interior. Washington, D.C.: U.S. Government Printing Office, 1978.

Smith, Baird M., and Frederec E. Kleyle, eds. "A Selected Bibliography on Energy Conservation in Historic Buildings." Technical Preservation Services Reading List. 1978. Rev. ed. Washington, D.C.: Technical Preservation Services Division, U.S. Department of the Interior, 1980.

Socolow, Robert H., ed. *Saving Energy in the Home: Princeton's Experiments at Twin Rivers.* Cambridge: Ballinger Publishing, 1978.

Stein, Richard G. *Architecture and Energy: Conserving Energy through Rational Design.* Garden City, N.Y.: Anchor Press, Doubleday, 1977.

Stein, Richard G., C. Stein, M. Buckley and M. Green, the Stein Partnership. *Handbook of Energy Use for Building Construction.* Washington, D.C.: U.S. Department of Energy, 1981. Available from National Technical Information Center.

Tatum, Rita. *The Alternative House: A Complete Guide to Building and Buying.* Danbury, N.H.: Reed Books, 1978.

U.S. Department of Housing and Urban Development. *In the Bank . . . Or Up the Chimney? A Dollars and Cents Guide to Energy-Saving Home Improvements.* Washington, D.C.: U.S. Government Printing Office, 1975.

U.S. Department of Housing and Urban Development, with U.S. Department of Energy. "Energy Conservation in Buildings." Bibliography DC-104. Rockville, Md.: National Solar Heating and Cooling Information Center, rev. periodically.

U.S. General Services Administration. *Energy Conservation Guidelines for Existing Office Buildings.* Washington, D.C.: Author, 1977.

Vila, Bob, with Jane Davison. "This Cold House." In *This Old House.* Boston: Little, Brown, 1980.

Wilson, Tom, ed. *Home Remedies: A Guidebook for Residential Retrofit.* From First National Retrofit Conference, sponsored by Mid-Atlantic Solar Energy Association with Center for Energy and Environmental Studies. Philadelphia: Mid-Atlantic Solar Energy Association (2233 Gray's Ferry Avenue, Philadelphia, Pa. 19146), 1981.

Wing, Charles. *From the Walls In.* Boston: Atlantic Monthly Press, Little, Brown, 1979.

Wolfe, Ralph, and Peter Clegg. *Home Energy for the Eighties.* Charlotte, Vt.: Garden Way Publishing, 1979.

Periodical Articles

Frederickson, John H. "Facilities Recycling for Energy Conservation." *National Association of Secondary School Principals Bulletin,* May 1979.

Green, Kevin W. "Energy Conscious Redesign" and "Uncommon Sense." *Research and Design, summer 1980.*

Labine, Clem. "The Energy-Efficient Old House." *Old-House Journal,* September 1977.

McClendon, Bruce W., and Ray Quay. "Targeted Energy Conservation Strategies in Galveston, Texas." *Urban Land,* June 1980.

"The Overselling of Insulation." *Consumer Reports,* February 1978.

Papian, William N. "Insulation and the Old House." *Old-House Journal,* August 1976, September 1976.

Peterson, Douglas C. "Ways to Save Energy and Stay Warm." *Historic Preservation,* March-April 1979.

Pilling, Ron. "Cooling the Natural Way." *Old-House Journal,* May 1979.

Progressive Architecture. Special energy issues. April 1980, April 1981.

Smith, Baird M. "National Benefits of Rehabilitating Existing Buildings." Supplement to *11593,* October 1977.

Tye, Ronald P. "Retrofit Thermal Insulation: An Evaluation of Materials for Energy Conservation." *Technology and Conservation,* fall 1979.

Solar Energy

American Planning Association. *Residential Solar Design Review: A Manual on Community Architectural Controls and Solar Energy Use.* Martin Jaffe and Duncan Erley. Prepared for U.S. Department of Housing and Urban Development, with U.S. Department of Energy. Washington, D.C.: U.S. Government Printing Office, 1980. Available from National Solar Heating and Cooling Information Center.

Anderson, Bruce, with Michael Riordan.

The Solar Home Book. Andover, Mass.: Brick House Publishing, 1976.

Anderson, Bruce, and Malcolm Wells. *Passive Solar Energy: The Homeowner's Guide to Natural Heating and Cooling.* Andover, Mass.: Brick House Publishing, 1981.

Barnaby, Charles S., Philip Caesar and Bruce Wilcox, The Berkeley Solar Group, with Lynn Nelson, Environ/Mental, consultant. *Solar for Your Present Home.* Sacramento: California Energy Resources Conservation and Development Commission (1111 Howe Avenue, Sacramento, Calif. 95825), 1978.

Butti, Ken, and John Perlin. *A Golden Thread: 2500 Years of Solar Architecture and Technology.* Cheshire Books. New York: Van Nostrand Reinhold, 1980.

Environmental Study Conference, U.S. Congress. *Guide to Solar Energy Programs.* Peter Rossbach and David Lauter, eds. Washington, D.C.: Author, 1980. Available from U.S. senators and representatives.

Feldman, Stephen L., and Robert M. Wirtshafter, eds. *On the Economics of Solar Energy.* Lexington, Mass.: D.C. Heath, 1980.

Hayes, Gail Boyer. *Solar Access Law: Protecting Access to Sunlight for Solar Energy Systems.* Cambridge: Ballinger Publishing, 1979.

Heschong, Lisa. *Thermal Delight in Architecture.* Cambridge: MIT Press, 1979.

Hill, James Edward. *Retrofitting a Residence for Solar Heating and Cooling: The Design and Construction of the System.* National Bureau of Standards. Washington, D.C.: U.S. Government Printing Office, 1975.

Hix, John. *The Glass House.* Cambridge: MIT Press, 1974.

Kreith, Frank, and Jan Kreider. *Principles of Solar Engineering.* New York: McGraw-Hill, 1978.

Lyons, Stephen, ed. *SUN! A Handbook for the Solar Decade.* San Francisco: Friends of the Earth, 1978.

Mazria, Edward. *The Passive Solar Energy Book: A Complete Guide to Passive Solar Home, Greenhouse and Building Design.* Emmaus, Pa.: Rodale Press, 1979.

McCullagh, James C., ed. *The Solar Greenhouse Book.* Emmaus, Pa.: Rodale Press, 1978.

National Solar Heating and Cooling Information Center. *Publications Catalog.* Rockville, Md.: Author, 1980.

Reif, Daniel K. *Solar Retrofit: Adding Solar to Your Home.* Andover, Mass.: Brick House Publishing, 1981.

Shurcliff, William A. *Superinsulated Houses and Double-Envelope Houses: A Survey of Principles and Practice.* Andover, Mass.: Brick House Publishing, 1981.

Solar Access Alliance. *A Model Ordinance to Establish a Minimum Energy Conservation Standard for Existing Residential Buildings.* Portland, Ore.: Author (P.O. Box 8210, Portland, Ore. 97207), 1981.

________________. *A Strategy to Provide and Protect Solar Access.* Portland, Ore.: Author, 1981.

Southern Solar Energy Center. *Passive Solar Retrofit Handbook.* Atlanta: Author, 1981.

Spetgang, I., and M. Wells. *Your Home's Solar Potential.* Barrington, N.J.: Edmund Scientific Company, 1976.

U.S. Department of Housing and Urban Development, with U.S. Department of Energy. "History of Solar Heating and Cooling." Bibliography DC-142. Rockville, Md.: National Solar Heating and Cooling Information Center, rev. periodically.

________________. "Reading List for Solar Energy." Bibliography DC-107. Rockville, Md.: National Solar Heating and Cooling Information Center, rev. periodically.

________________. "Solar Retrofit Bibliography." Bibliography DC-121. Rockville, Md.: National Solar Heating and Cooling Information Center, rev. periodically.

Periodical Articles

AIA Research Corporation. "Solar Architecture." *Research and Design,* January 1978.

Dubin, Fred S. "Solar Energy Design for Existing Buildings." *ASHRAE Journal,* November 1975.

Green, Kevin, W. "Passive Cooling." *Research and Design,* fall 1979.

Museums and Historic Sites

Ambrosino, P.E. "Energy Conservation at the Cooper-Hewitt Museum: Renovation/Retrofitting for Improved Mechanical System Efficiency." *Technology and Conservation,* spring 1979.

"Energy: Historical Agencies Make a Special Case." *History News,* May 1980.

Heritage Conservation and Recreation Service, U.S. Department of the Interior. *Rehabilitation: Claremont 1978. Planning for Adaptive Use and Energy Conservation in an Historic Mill Village.* Washington, D.C.: U.S. Government Printing Office, 1979.

Matthai, Robert A. "Energy Conservation and Management: A Critical Challenge for Cultural Institutions." *Technology and Conservation,* spring 1978.

Matthai, Robert A., ed. *Energy and the Cultural Community.* Final Report of the Arts-Energy Study to the National Endowment for the Arts. Flushing, N.Y.: Energy Information Clearinghouse for the Cultural Community, 1979.

__________________. *Energy Management for Museums and Historical Societies.* Flushing, N.Y.: Energy Information Clearinghouse for the Cultural Community, 1981.

Miller, Hugh. "Energy Conservation and Historic Sites: Old Buildings, New Tools." *Energy Ideas,* January-February 1978. (Energy Conservation Project, National Recreation and Park Association).

__________________. "Energy Management Planning: A Systematic Approach for Museums and Historic Buildings." *Technology and Conservation,* fall 1979.

Sources of Energy Information

Advisory Council on Historic Preservation
1522 K Street, N.W., Suite 430
Washington, D.C. 20005

As part of its responsibilities to advise the president and Congress on federal preservation policy and to protect historic sites, sponsored the study *Assessing the Energy Conservation Benefits of Historic Preservation.*

AIA Research Corporation
1735 New York Avenue, N.W.
Washington, D.C. 20006

Nonprofit organization created by the American Institute of Architects, but operating separately from it, to research issues of national significance affecting the built environment. The corporation's applied research has explored such subjects as energy conservation, solar energy, seismic design and related architectural problems. It formerly published *Research and Design.*

American Section of the International Solar Energy Society
Research Institute for Advanced Technology
U.S. Highway 190 West
Killeen, Tex. 76541

State and regional chapters provide professional help, publications and audiovisual materials on solar energy.

Energy Information Clearinghouse for the Cultural Community
New York Hall of Science
Box 1032
Flushing, N.Y. 11352

Cooperative program of museum and cultural service organizations to provide information, publications and expert advice on energy concerns of cultural institutions.

Center for Advanced Computation
Energy Library
University of Illinois
Urbana, Ill. 61801

Has conducted innovative research into the energy utilization and embodiment of buildings. Its *Energy Research Group Abstract List* presents the center's research on energy supply, statistics, policy, recycling and related topics.

Citizens' Energy Project
1110 Sixth Street, N.W., Suite 300
Washington, D.C. 20001

Offers research and educational materials and a state-by-state citizens energy directory listing organizations, individuals and companies providing energy services.

Golden Gate Energy Center
Golden Gate National Recreation Area
Building 1055, Fort Cronkhite
Sausalito, Calif. 94965

Private, nonprofit organization established as a demonstration center for conservation and renewable energy technologies. The center, located in a converted military site, provides educational, workshop and conference programs.

Historic House Association of America
1600 H Street, N.W.
Washington, D.C. 20006

Provides advice, publications and conferences on historic house maintenance and restoration concerns, including energy conservation.

Manchester Citizens Corporation
1120 Pennsylvania Avenue
Pittsburgh, Pa. 15233

Private, nonprofit community organization that is directing revitalization of the historic Manchester neighborhood in Pittsburgh, including implementation of energy guidelines developed under a Carnegie-Mellon University—U.S. Department of Energy demonstration project.

National Association of Counties Energy Project
1735 New York Avenue, N.W.
Washington, D.C. 20006

Provides information and technical assistance to 2,000 member counties to assist them in responding to energy issues such as conservation, development of renewable energy sources, contingency planning and managing the impact of energy development.

National Center for Appropriate Technology
P.O. Box 3838
Butte, Mont. 59701

Provides grants and technical assistance to low-income individuals and groups for small, innovative conservation and solar energy projects; also issues related publications.

National Solar Heating and Cooling Information Center
P.O. Box 1607
Rockville, Md. 20850
(800) 523-2929
(800) 500-4700 (Alaska and Hawaii)
(800) 462-4983 (Pennsylvania)

Established by HUD in cooperation with DOE (operated by the Franklin Research Center) to provide information, publications, demonstration data, speakers and listings of conferences and workshops. It is the basic information center on all aspects of solar heating and cooling, technical and nontechnical, domestic and foreign.

National Trust for Historic Preservation
1785 Massachusetts Avenue, N.W.
Washington, D.C. 20036

Maintains information on energy conservation issues affecting historic buildings, including design review by historic district commissions, and plans demonstration conservation strategies for its historic properties. Sponsored public information programs and a symposium on energy conservation through preservation during Preservation Week 1980.

Solar Energy Industries Association
1001 Connecticut Avenue, N.W., Suite 800
Washington, D.C. 20036

Trade organization of manufacturers, distributors and designers of solar energy equipment; publishes a newsletter and directories.

Solar Energy Institute of North America
1110 Sixth Street, N.W.
Washington, D.C. 20001

Organization of solar professionals, educators and consumers established to promote the use of solar systems.

Solar Energy Research Institute
1617 Cole Boulevard
Golden, Colo. 80401

Supports solar research, development and demonstration activities of DOE; also maintains a library and provides information to state energy offices.

Solar Lobby
1001 Connecticut Avenue, N.W.
Washington, D.C. 20036

Provides information and publications on public solar issues, including data on state solar legislation.

State Energy Offices

Services vary in each state from development of state energy policy to provision of assistance to community groups such as educational materials, information on local organizations, technical aid and professional services.

U.S. Department of Energy
Washington, D.C. 20585

Administers various conservation programs from Washington and 10 regional offices in Boston, New York City, Philadelphia, Atlanta, Chicago, Dallas, Kansas City, Mo., Lakewood, Colo., San Francisco and Seattle; participates in cooperative programs with other federal agencies; sponsors four regional information centers that distribute solar energy information tailored to each region:

Regional Solar Energy Centers

Northeast Solar Energy Center
470 Atlantic Avenue
Boston, Mass. 02110

Southern Solar Energy Center
61 Perimeter Park
Atlanta, Ga. 30341

Mid-American Solar Energy Complex
8140 26th Street
Bloomington, Minn. 55420

Western Sun
715 S.W. Morrison
Portland, Ore. 97205

U.S. Department of Housing and Urban Development
451 7th Street, S.W.
Washington, D.C. 20410

Provides programs designed to establish and expand residential energy conservation activity. Information is available from Washington and regional HUD offices in Boston, New York City, Philadelphia, Atlanta, Chicago, Fort Worth, Kansas City, Mo., Denver, San Francisco and Seattle, as well as area offices in each state.

U.S. Department of the Interior
National Park Service
Technical Preservation Services Division
Washington, D.C. 20240

Develops and publishes a wide range of technical information about techniques for preserving and maintaining historic properties, in addition to related federal preservation program activities. Has published on energy conservation issues and is continuing research in this area.

U.S. Internal Revenue Service
Office of the Chief Counsel
1111 Constitution Avenue, N.W.
Washington, D.C. 20224

Adminsters tax credit provisions of the Internal Revenue Code encouraging energy conservation and renewable energy applications. IRS publications and forms are available to local IRS offices.

Specialized Booksellers

National Bureau of Standards
Technical Information and Publications Division
U.S. Department of Commerce
Gaithersburg, Md. 20760

National Technical Information Center
U.S. Department of Commerce
5285 Port Royal Road
Springfield, Va. 22161

Preservation Bookshop
National Trust for Historic Preservation
1600 H Street, N.W.
Washington, D.C. 20006

Superintendent of Documents
U.S. Government Printing Office
Washington, D.C. 20402

The Contributors

Michael L. Ainslie is president of the National Trust for Historic Preservation. Before assuming this position in July 1980, he was senior vice president and chief operating officer of N-ReN Corporation in Cincinnati. From 1969 to 1970 Ainslie was deputy director of the New York City Model Cities Program. He also has served with Sea Pines Company, Inc., and has overseen the preservation and restoration of numerous historic buildings in Savannah, Ga., and Cincinnati.

Calvin W. Carter is a member of the Advisory Council on Historic Preservation and chaired the committee responsible for the council study *Assessing the Energy Conservation Benefits of Historic Preservation: Methods and Examples.* He is a member of the Florida Historic Preservation Review Council, the advisory board of Tampa Preservation and the Arts Council of Tampa-Hillsborough County; an adviser to the Neighborhood Rehabilitation Council; and past president of the board of trustees of the Tampa Bay Arts Center. Carter is president and cheif executive officer of the O.H. Carter Company, Tampa, Fla.

Frank B. Gilbert is chief counsel for landmarks and preservation law, National Trust for Historic Preservation. From 1965 to 1974 he served as secretary and then executive director of the New York City Landmarks Preservation Commission. For more than 10 years he was involved in the successful effort to save Grand Central Terminal in New York.

Paul Goldberger, the *New York Times* architecture critic, writes widely on architecture, urban planning, preservation and related design issues. He is a contributing editor of *Portfolio* and has written for *Architectural Record, Progressive Architecture, Architectural Forum, New York, New York Review of Books, Times Literary Supplement, Art News* and *Esquire.* Goldberger is the author of *The City Observed. New York: An Architectural Guide to Manhattan* (Random House, 1979) and contributions to a variety of books on architectural subjects, including *Old and New Architecture: Design Relationship* (Preservation Press, 1980).

Gary Long, AIA, is director of architecture, University of Colorado. He also is a partner in Long Hoeft Architects, a preservation practice in Denver. Long maintains an interest in energy-conscious architecture, old and new, and is coauthor of *Energy Conservation Standards* (McGraw-Hill, 1978).

Steve Mooney is program coordinator and energy system specialist for the Golden Gate Energy Center. He served formerly as technical writer and communications specialist for solar technologies with the Solar Energy Research Institute in Golden, Colo. Recent experience includes research for the U.S. Department of Energy-sponsored Hamilton Air Force Base-Marin Solar Village Project and program coordination of a municipal energy planning project for Marin County, Calif.

James S. Moore, Jr., P.E., is a project manager with Mueller Associates, Inc., consulting engineers, Baltimore, Md. For the past seven years, he has participated in a variety of energy conservation studies and design projects addressing both new and existing residenial and commercial buildings. Moore also has participated in solar energy system design and analysis projects for a spectrum of building types, including cost and performance review surveys of numerous solar installations. His paper is based on remarks presented to the 1980 annual conference of the Maryland Historical Trust in Baltimore.

Nathaniel Palmer Neblett, AIA, Alexandria, Va., is an architect specializing in the preservation and restoration of old buildings. He served previously as historical architect for the National Trust for Historic Preservation. His paper is adapted from Neblett's *Energy Conservation in Historic Homes* (Historic House Association of America, 1980).

Web Otis has been executive director of the Golden Gate Energy Center since June 1980. He served previously as director of the Office of Invention and Small Scale Technology under the Assistant Secretary for Solar Energy, U.S. Department of Energy. Otis also has implemented pilot appropriate technology grants for the San Francisco Regional Office of DOE and served as a special assistant to the Secretary of the Interior. Otis is a past chairman of the Federal Regional Council, San Francisco.

Neal R. Peirce is a founder and contributing editor of the *National Journal,* Washington, D.C. His column on urban issues is syndicated nationally by the Washington Post Writers Group. Peirce is the author of several books, including *The Megastates of America* and eight others in a series on "People, Politics and Power" in the 50 states (W.W. Norton, 1972-80). Peirce's observations are adapted from remarks he made to the 1980 annual conference of the Maryland Historical Trust in Baltimore.

Douglas C. Peterson is president of Nordic Insulation, Inc., Brockton, Mass., a manufacturer and distributor of insulation and related materials. He was formerly general manager of The Energy Bank, a division of Technology + Economics, Inc., a consulting firm in Cambridge, Mass., that conducts energy audits of houses and advises on energy-related improvements. His chart appeared originally in the March-April 1979 issue of *Historic Preservation,* the magazine of the National Trust.

Travis L. Price III is a principal architect with Price and Partners Architects in Takoma Park, Md., and a visiting lecturer at the University of Maryland School of Architecture. He has designed and built innovative energy projects throughout the United States during the past decade, including solar applications in historic adobes in New Mexico, solar and wind systems in New York City and several solar projects for the Tennessee Valley Authority. Price is project manager for the Carnegie-Mellon University study of the Manchester neighborhood described here. Associates on this project include Prof. Volker Hartkopf, principal investigator for the case study; Richard Glance, preservation architect; Stanley Lowe, executive director of the Manchester Citizens Corporation; Susan Cichetti, Sheffield block development manager for MCC; Rosalind Wooten, community energy outreach coordinator for MCC; Thomas Cox, financial consultant, Urban Design Associates (new construction architects); Steve Lee, project architect; and Carl Detwiler, restoration architect.

Fredric L. Quivik is building recycling specialist at the National Center for Appropriate Technology, Butte, Mont. A graduate of the University of Minnesota Architecture School and the master's degree program in historic preservation at Columbia University, he has served also as a historian for the National Architecture and Engineering Record, U.S. Department of the Interior. During summer 1979 he was senior architect on a HAER Rehab Action Team in Butte, and in 1980 he completed a HAER inventory of historic bridges throughout Montana. Quivik also is an instructor at the Montana College of Mineral Science and Technology in Butte.

Theodore Anton Sande is executive director of the Western Reserve Historical Society in Cleveland. He also is chairman of the board of the International Committee for the Conservation of the Industrial Heritage and is a founder and served as first president of the Society for Industrial Archeology. Sande is a former vice president for historic properties of the National Trust for Historic Preservation and a former member of the Old Georgetown Board of the U.S. Commission of Fine Arts. He is the author of *Industrial Archeology: A New Look at the American Heritage* (Stephen Green Press, 1976; Penguin, 1978), and has taught at Williams College and the University of Pennsylvania.

John C. Sawhill was deputy secretary, U.S. Department of Energy, at the time his remarks were prepared. He subsequently was appointed chairman of the U.S. Synthetic Fuels Corporation. His previous responsibilities include service as president of New York University; administrator of the Federal Energy Administration; associate director for national resources, Office of Management and Budget; and senior vice president, Commercial Credit Corporation. He served previously on the boards of directors of the American Association for the Advancement of Science and Common Cause. Sawhill is coauthor of *Nuclear Power Issues and Choices: Report of the Nuclear Energy Policy Study Group* (Ballinger, 1977) and principal author of *Energy: Managing the Transition, Energy Conservation and Public Policy* (Prentice Hall, 1979) and *Improving the Energy Efficiency in the American Economy.*

Ann Webster Smith is deputy commissioner for historic preservation, New York State Office of Parks and Recreation. From 1975 to 1979, she was deputy to the Secretary General of the International Council on Monuments and Sites in Paris. Smith also has served as director of compliance and intergovernmental planning for the Advisory Council on Historic Preservation and in the U.S. Department of Transportation's Office of Environment and Urban Systems.

Baird M. Smith, AIA, is chief of the Branch of Preservation Technology, Technical Preservation Services Division, National Park Service, U.S. Department of the Interior, where he has been on the staff since 1976. He currently directs the division technical publications program and research projects in the areas of energy conservation, solar energy applications, stone conservation and seismic reinforcement of historic buildings. Smith worked previously with the National Bureau of Standards and the National Trust for Historic Preservation. He holds degrees in architecture and historic preservation.

James Vaseff, AIA, is acting chief, Preservation Services Division, Atlanta regional office, National Park Service, U.S. Department of the Interior. Before assuming this position, he was acting principal architect of the Historic American Engineering Record. Vaseff supervised HAER rehabilitation action teams in North Carolina and Danville, Va., before moving to HAER's national demonstration program. He has practiced architecture in Boston, England and North Carolina and taught at the College of Architecture, University of North Carolina.

William I. Whiddon is a senior associate with Booz, Allen and Hamilton, Inc., Bethesda, Md. He specializes in the energy-conscious design of commercial and residential buildings and the commercialization of new energy technologies. Whiddon is the primary author of the Advisory Council on Historic Preservation report *Assessing the Energy Conservation Benefits of Historic Preservation: Methods and Examples* and is coauthor of *People, Energy and Buildings: A Graphic Approach to Energy Conscious Design,* a 1978 Booz, Allen report for the Department of Energy's Office of Federal Programs, Conservation and Solar Applications, and *Energy Graphics* (Booz, Allen, 1980). He holds degrees in architecture and aeronautical engineering and has taught at the College of Architecture and Design, Kansas State University.

Index

Buildings are listed under their city locations. Page numbers in italics refer to illustrations or their captions.